SpringerBriefs in Physics

For further volumes:
http://www.springer.com/series/8902

Norbert Dragon

The Geometry of
Special Relativity—
a Concise Course

Norbert Dragon
Institut für Theoretische Physik
Leibniz Universität Hannover
Hannover
Germany

ISSN 2191-5423   ISSN 2191-5431   (electronic)
ISBN 978-3-642-28328-4   ISBN 978-3-642-28329-1   (eBook)
DOI 10.1007/978-3-642-28329-1
Springer Heidelberg New York Dordrecht London

Library of Congress Control Number: 2012938635

Printed on acid-free paper

Springer is part of Springer Science+Business Media (www.springer.com)

Preface

So einfach wie möglich, aber nicht einfacher

As simple as possible, but not simpler. This guideline of Albert Einstein obliges in particular each presentation of relativistic physics, a subject which often puzzles laymen, stirs their imagination, and tantalizes their comprehension, unnecessarily, because relativistic physics relies on simple geometric notions.

If one wants to understand the basic features of the theory of relativity then one does not need coordinates or virtual systems of clocks, which fill the universe, no more than millimeter paper and coordinate axes are required for Euclidean geometry. One only has to consider what observers see rather than to argue that this or that observer is right. Relativity is a physical, not a judicial theory.

The slowdown of moving clocks and the shortening of a moving measuring rod unfold naturally from the principle of relativity, just as a tilted ladder is less high than an upright ladder. Clocks are no more mysterious than mileage meters, and show a distance between start and end which depends on the way in between. This is the unspectacular answer to the seemingly paradoxical aging of twins. Just as no one is puzzled by a triangle, where the straight line between two edges is shorter than the detour over the third edge, no one should be shocked by the conclusion and experimental verification, that a clock picks up more time on a straight history as compared to the twin clock of a traveler who takes a detour.

The first two chapters are intended to be understandable in essence also to non-physicists with little mathematical knowledge. Their simplicity, however, may be deceptive. Real understanding requires careful consideration of the arguments, the equations, and the diagrams, preferably by reading equipped with a pencil and paper.

The following chapters presume mathematical knowledge which physicists and mathematicians acquire during their undergraduate years. To clarify more complicated questions we introduce coordinates as functions of the measured times and directions of light rays and deduce the Lorentz transformations which relate these values to the ones which moving observers measure. These transformations determine how velocities combine, what pictures are seen by moving observers, and how the energy and momentum of a particle depend on its velocity.

Chapter 4 assembles the basics of mechanics and applies them to relativistic particles. Stress is laid on the correspondence between physics and geometry, between conserved quantities like energy, momentum and angular momentum, and symmetries like a shift in time or space or a rotation or a Lorentz transformation. Jet spaces, which are introduced and used in this investigation, may strike the reader as an unnecessary complication. But they provide the clearest and therefore simplest setting to exhibit the correspondence of conserved quantities and infinitesimal symmetries.

Chapter 5 presents electrodynamics as a relativistic field theory and in particular shows that changes of the electric charges cause changes of the electromagnetic fields with the speed of light. The electrodynamic interactions are invariant under dilations, which is why they cannot explain the particular values of particle masses or the particular sizes of atoms.

In the last chapter we discuss the mathematical properties of the Lorentz group. It acts on the directions of light rays just as the Möbius transformations act on the Riemann sphere.

The text originated from courses which I taught on the subject and from my answers to questions which were frequently asked in the newsgroup de.sci.physik. After a few years the notes changed nearly no more and slumbered on my homepage with a few hundred interested visitors per year until Christian Caron from Springer Verlag encouraged me to have them published. Whether this kiss of a prince awoke a sleeping beauty or a frog, still to be thrown against the wall, is the reader to judge.

Helpful comments and patient listening were contributed by Frédéric Arenou, Werner Benger, Christian Böhmer, Christoph Dehne, Jürgen Ehlers, Christopher Eltschka, Chris Hillman, Olaf Lechtenfeld, Volker Perlick, Markus Pössel, and Bernd Schmidt. Ulrich Theis translated the early versions of the notes. Sincere thanks are given to Ulla and Hermann Nicolai for their friendly hospitality during my stay at the Albert-Einstein-Institut der Max-Planck-Gesellschaft.

Hannover, Germany, January 2012   Norbert Dragon

Contents

Chapter 1
Structures of Spacetime

Abstract Simple geometric properties of spacetime and free particles underlie the theory of relativity just as Euclidean geometry follows from simple properties of points and straight lines. The vacuum, the empty four-dimensional curved spacetime, determines straight lines and light rays. In the absence of gravity, the vacuum is isotropic and homogeneous and does not allow to distinguish rest from uniform motion. Therefore, contrary to Newton's opinion, the vacuum cannot contain the information about a universal time which could be attributed to events. Whether two different events are simultaneous depends on the observer—just as in Euclidean geometry it depends on a given direction whether two points lie on an orthogonal line.

1.1 Properties of the Vacuum

We can denote a point in space by specifying how far away it is ahead, to the right and to the top of a chosen reference point. These specifications are called coordinates of the point. One needs three coordinates in order to specify any one point. Space is three-dimensional. The coordinates of a point depend of course on the choice of the reference point and on which directions the observer chooses as ahead, right and above.

As for appointments in daily life, for physical processes not only the position is important, where an event takes place, but also the time when it occurs. The set of all events, spacetime, is four-dimensional, because to specify a single event one needs four labels, the position where it takes place and the time when it occurs. The position and time specifications which label an event depend—just as the three coordinates of a position—on the observer.

The four-dimensional spacetime fascinates and beats our imagination which is trained in everyday life. Nevertheless it is quite simple. We can easily envisage a stack of pictures, as they are stored in a film reel, which show the sequel of three-dimensional situations. Thereby one conceives the four-dimensional space-

N. Dragon, *The Geometry of Special Relativity—a Concise Course*,
SpringerBriefs in Physics, DOI: 10.1007/978-3-642-28329-1_1,
© The Author(s) 2012

time the same way as an architect, who draws two-dimensional blueprints, horizontal plans and transversal sections to envisage a three-dimensional building.

Using the same means we depict the sequel of events in two-dimensional space-time diagrams. For example, the geometric figure of two intersecting straight lines shows the physical process that two particles move uniformly and collide in the event, where the lines intersect. If one would display only the position and not the time one would not know whether the particles pass the same position at the same time or fail to meet each other.

The physical findings add the insight, which is alien to our intuition, that the four-dimensional spacetime is an entity which is decomposed into layers of equal time only by the observer. Different from what Newton thought, these slices of equal time do not coincide for observers which move relative to each other.

That simultaneity depends on the observer is the largest obstacle for understanding relativity. Not only the three coordinates of the position, but also the time which denotes an event, depend on the observer. Spacetime has no measurable universal time which pertains to the events.

Each event E determines the later events which can be influenced by E by means of light or electromagnetic signals, and vice versa, it determines the earlier events from which it could have been influenced. These events form the forward and backward lightcone of E, which both belong geometrically to each event in spacetime.

An unaccelerated clock which passes two events shows the time which passes in between. This time does not depend on the particular type of clock and is a geometrical property of the two events, their temporal distance. From this time all length standards derive, in particular spatial distance is the time which it takes light to run back and forth (1.5). The temporal distance of events imparts a geometry to spacetime which is similar to Euclidean geometry in many respects. As we will see, the events with the same temporal distance from a chosen event lie on a hyperboloid and not, as in Euclidean geometry, on a sphere. The geometry of spacetime is ordinary school geometry, with circles replaced by hyperbolas.

To avoid the multitude of possible effects, we investigate processes in an empty region of spacetime, the vacuum, from which all particles have been removed and all influences from outside, such as electric and magnetic fields, are shielded. This vacuum is the stage on which we study the behavior of light and particles that are seen by observers and are measured with clocks and measuring rods.

As simple as the idea of a vacuum seems, it is an idealization and can be realized only approximately. We are constantly passed by neutrinos, which come from the sun and single out a particular direction in the otherwise isotropic space. We cannot shield our experiments from these neutrinos since they do not interact sufficiently. But since neutrinos penetrate everything, they do not bother either and cause effects only if we look for them on purpose.

The cosmos filled by background radiation is not a vacuum. This radiation is a remnant from the early evolution of the universe and defines a rest frame through which the sun moves with a speed of roughly 370 km/s [28]. This background radiation can be shielded by walls, but the walls have to be cooled so that the heat radiation of the walls does not fill the space.

Fig. 1.1 Orbits around the
Earth

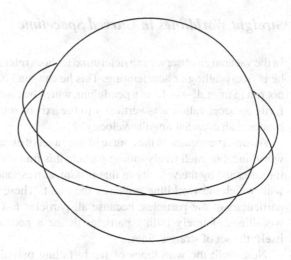

Omnipresence of Gravity

Even after removing all particles and shielding from all external influences the vacuum retains structure.

Gravity cannot be shielded or extracted from the vacuum. It therefore belongs to the properties of the empty spacetime, though in special arrangements one can compensate gravitational attraction in a region by additional masses. For example, if one completes a segment of a spherical shell, which causes gravitational attraction, to a complete shell, then in the interior of the shell the gravitation of the segment is compensated. However, one cannot compensate unforeseen disturbances from outside by a Faraday cage with particles, which can move freely in a wire with negligible inertia. For freely moving particles, the state with lowest energy is not a vanishing gravitational field, but, because gravity is attractive, the closest packing of the particles and the largest possible gravitational attraction. They do not shield gravity but enlarge it. Also, there are no bodies which are inert and nearly insensitive to gravity. Irrespective of their masses all test particles fall in the same way.

One cannot completely transform away gravity by performing experiments in freely falling laboratories. In a laboratory which orbits the earth in free fall one can distinguish three different directions, behind, below and beneath, by gravitational effects without view to the earth and without reference to the walls. Test particles in the laboratory orbit the earth in ellipses, in the simplest case in circles. If the test particles orbit circles in the same plane with different radii, then the particle which is nearer to earth is faster and departs from the other. If the test particles cycle behind each other in the same circle, then their distance remains unchanged. If they orbit initially beneath each other in different circles of the same radius, then the orbital planes intersect and the circles intersect twice per revolution: freely falling particles beneath each other oscillate around each other with the orbital frequency.

Straight Worldlines in Curved Spacetime

In the vacuum an observer can determine without reference to other positions whether he is freely falling or accelerating. This he can read off from an hour glass—it does not run in free fall—or from a pendulum, which he carries along. If it swings back and forth, an acceleration acts vertically to the axis of rotation, otherwise the pendulum rotates with constant angular velocity.

An observer passes in the course of time a set of events. This line in spacetime is his worldline. For each freely falling particle this line is determined by an event, which it passes and by the velocity in that instant, corresponding to a point and a direction with which the worldline traverses the point. These worldlines do not depend on particulars of the particles, because all particles fall the same way. Therefore the worldlines of freely falling particles define a geometrical structure of spacetime itself: the set of straight lines.

Note well: the worldlines of freely falling particles and of flashes of light are the straight lines of the four-dimensional spacetime, but their spatial projection are not straight lines of the three-dimensional space. Through each point in space and with given direction there pass different parabolas which are traced out by falling particles with different velocities. These curves in three-dimensional space are not determined—differently from what one has to require from straight lines—by a point and a direction.

Among the spatial curves of freely falling particles one can and does choose a class of curves to define straight lines, if gravity does not change with time. Straight is the path of light. Whether an edge is straight is checked by comparing with light, by sighting along the edge. But light rays are gravitationally deflected and can intersect each other repeatedly. They do not satisfy Euclid's axiom of parallels. They define straight lines in a space which is curved by gravity.

If gravity changes in time because the masses move which generate gravity, then the light rays do no longer define straight spatial lines, because the ways to and fro differ.

One could imagine to attribute the label straight to other lines in spacetime, for example to lines which in some coordinate system can be drawn with a ruler. But these lines are no property of spacetime and test particles traverse them only if they are subject to forces which differ for different particles. The only worldlines which are singled out by nature are the worldlines of freely falling particles, including the worldlines of flashes of light.

If one follows mentally the path of freely falling particles, which in Fig. 1.1 orbit the earth in different circles, then one realizes that spacetime is curved. If one relates the positions of the second particle to the positions of the first particle, then the straight worldline of the second particle oscillates around the straight worldline of the first particle, straight lines can intersect each other repeatedly.

The reason for the curvature of spacetime and the relative motion of freely falling particles is the fact, that gravity is not the same everywhere: the attraction is stronger the nearer the particles and in equal distance it acts in different directions. By the

Fig. 1.2 Straight lines in curved spacetime

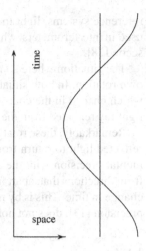

different gravitational attraction one can distinguish different positions and directions. If however, one restricts physical evolutions to such short times and small regions that the inconstancy of gravity does not make itself felt at the given precision of measurement, then the effects of gravity become imperceptible in a freely falling system of reference. In sufficiently small regions the curvature of spacetime is insensible and spacetime has the geometric properties of a flat space (Fig. 1.2).

We cannot shield gravity but want to avoid the related complications of a curved spacetime. Therefore we restrict our considerations to short times and distances such that gravitational effects are immeasurably small or we account mentally for the known gravitational effects and subtract them from the observed behavior of the physical systems.

If then, one has shielded all external influences and subtracted gravitational effects, then one cannot measure the time and the position of an event without reference to other events—just as little as at sea one cannot measure latitude without reference to the sun and time in Greenwich or without GPS. Physical evolutions are the same everywhere and at any time: the spacetime is homogeneous. Similarly physical evolutions are the same in all directions: the spacetime is isotropic.

Rotational Motion

Rotational motion, the temporal change of directions, can be measured—different from uniform straight motion—without reference to other bodies such as the distant stars. If one rotates and emits light into some direction, then the reflected light is seen to return from a different direction [32]. In a rotating cinema one projects into one direction and observes from a different one. Only for nonrotating observers does reflected light return from the direction into which it had been emitted. In rotating reference systems, light to and fro does not follow the same path. This property is used in interferometers, which measure rotation with a precision of 10^{-8} degrees per second [8].

The situation where some object orbits around the observer is different from his own rotation. In both situations the light rays from the object come from directions which change in the course of time, but if one does not rotate one sees each single light ray reflected from the object return from the direction into which it was sent.

Remarkably, these rotation-free reference frames defined by the local property of reflected light to return from the direction of emission coincide within high experimental precision with the systems in which the light from the distant stars comes from directions that, apart from parallax, aberration and the motion of the star, do not change in time. This is by no means self-evident and, measured with today's highest precision [13], does not hold if one orbits the rotating earth.

Lightcone

Long before Einstein's theory of relativity the principle of relativity was known, that one cannot distinguish by any effect of Newtonian mechanics whether an unaccelerated observer moves or rests.

However, one was convinced that this principle of relativity is valid only approximately because it was known since 1676 from Olaf Rømer's observation and interpretation of the orbital periods of the four large moons of Jupiter, Jo, Europa, Ganymed and Kallisto, that c, the speed of light in the vacuum, is finite. Therefore light was assumed to single out an absolute rest system, in which light propagated equally fast in all directions and in which its medium, the ether, rested. For an observer, moving with a velocity v with respect to the ether, light should propagate in different directions with different velocities ranging from $c - v$ to $c + v$.

This conclusion is obvious and *wrong*: no experiment has ever measured that a moving source of light emits light which in the vacuum propagates with different velocities in different directions. Moreover, never has the motion of the observer made him to register in the vacuum different velocities of light in different directions. This is the result of the seminal experiment of Albert Michelson.[1] The most astonishing property of the ether is that never one found a trace of it. Ether has all the properties of the vacuum, it is the vacuum. Light propagates in the vacuum with a velocity which does not allow to distinguish a stationary observer from a uniformly moving observer.

Principle of Relativity: *The speed of light in the vacuum does not depend on the motion of the source. No physical observation allows to distinguish an observer at rest from an uniformly moving observer.*

In the vacuum there is no faster or slower light. Light does not outrun light [7].

For example, in 1987, one observed a supernova in the Large Magellanic Cloud, SN 1987a, which exploded 160000 years ago and where the luminous plasma was

---

[1] Experimental findings are discussed in detail in [34].

emitted in all directions with a velocity of 25 000 km/s. If this velocity v had added to the velocity of light to $c' = c + v$, then the light from the plasma which moved towards us would have arrived 12 000 years earlier than the light from the plasma which moved transversal to the line of sight.

Nobody had observed the star when the first light of the explosion arrived here, but in the explosion also neutrinos were emitted whose time of arrival was recorded. As one later found, they had triggered the counters one hour before one looked into the direction of the star and saw the explosion. At that time it was completely visible. Thus, there could have been runtime differences of at most one hour for the different lightrays.

A year has roughly $365 \cdot 24$ hours. With a runtime of 160 000 years the velocities of the light rays therefore were equal up to $1/(160\,000 \cdot 365 \cdot 24) \approx 0,7 \cdot 10^{-9}$, i.e. in the first nine decimals.

By the way, this observation implies also [31] that the neutrinos had moved with the speed of light within this precision, and that their mass is less than 10 eV/c^2.

That the speed of light in the vacuum is independent of the speed of the source agrees with the conclusions which one can draw from Maxwell's equations (5.3, 5.4) for the electromagnetic fields. From these equations we deduce (page 97) that a charge which is at the position $\mathbf{x}'$ at the time t' influences the electric and magnetic fields at the position $\mathbf{x}$ at the time t,

$$c\,(t - t') = |\mathbf{x} - \mathbf{x}'|\,, \tag{1.1}$$

which is later than t' by the runtime of light $|\mathbf{x} - \mathbf{x}'|/c$. The time t does not depend on the velocity of the charge, which causes the field.

The events $(t, \mathbf{x})$ constitute the forward lightcone of the event $(t', \mathbf{x}')$; electromagnetic causes produce effects with the speed of light in the vacuum.

The independence of the propagation of light from the velocity of the source does not imply that other properties, such as the color of the light, the direction of the light rays and the intensity of the radiation, do not depend thereon. The direction of the incoming light rays, the color of the light and the number of photons per time and unit area, the luminosity, do depend on the velocity of the source and the velocity of the receiver. The intensity of electromagnetic radiation depends on the acceleration of the charges that emit the radiation.

The independence of the propagation of light from the velocity of the source is illustrated in the spacetime Fig. 1.3. A stationary observer traverses the straight worldline $\mathscr{O}_0$, his position x remains unchanged at all times t; a second observer traverses the worldline $\mathscr{O}_v$ and moves uniformly into x-direction. If both observers are at some time at the same position and emit a flash of light, then the light propagates from this event E equally fast in all directions[2] and independently of whether the light source moves.

---

[2] Our diagrams show only one spatial dimension. Therefore there is only the direction forwards or backwards.

Fig. 1.3 Observers with
outgoing light rays

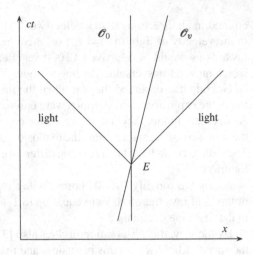

We call the worldline of a flash of light "light ray". Our diagrams are rotated such that the light rays which are emitted from E in forward and backward direction are traversed with increasing time from the bottom to the top of the diagram symmetrically to the vertical axis. We choose the units such that light rays include an angle of $\pm 45°$ with the axes. The light rays which pass through other events E' are parallel to the light rays through E, because light does not outrun light. Light rays in the same direction do not intersect.

Each event E determines the later events, which can be influenced from E by light and conversely it determines the earlier events which could have influenced E. These events constitute the forward and backward lightcone of E. The lightcone of each event does not depend on the motion of the source nor, for instance, on the color of light. The lightcones belong to spacetime itself, like the groove to a gramophone record. They are a geometrical structure of spacetime.

Apart from the light rays there are no other distinguished straight lines in the empty spacetime. There is no ether which traverses detectable worldlines and there is no measurable universal time which would define layers of equal time. The time- and space-axes are, different from lightcones, no geometric structure of spacetime but depend on the observer. Therefore and in order to not overcrowd the diagrams we leave the axes away—the geometry of spacetime is independent of coordinate axes just as Euclidean geometry.

If light and electromagnetic waves were the oscillations of an ether which filled the vacuum, such as sound is an oscillation of air and other matter, and if the worldlines were detectable, which are traversed by the constituents of the ether, then one could measure the motion relative to the ether and could distinguish it from rest.

Could one measure a universal time for each event with a result which is the same for all observers and which therefore pertained to the event itself, then one could distinguish an observer at rest from an observer in uniform motion by the physical property, that flashes of light, which he emitted in one event into different directions and which are reflected at the same universal time would return to him, but not to the moving observer, in the same instant.

The experimental results that in the vacuum one cannot distinguish rest from uniform motion therefore states that no ether is detectable and that no universal time is measurable. This disproves Newtonian ideas and seems to contradict common sense, our everyday experience. But our experience is restricted to low velocities and does not know the vacuum. How nature behaves beyond our routine environment is clarified by physicists by observations and experiments.

1.2 Measuring Rods

Metersticks are subject to many influences which restrict their precision. The length of metersticks varies with their temperature. On earth one has to see, i.e. to examine with light, whether they are bent by their weight. It causes them to be shorter if they stand and to be longer if they hang. Over larger distances there are no rigid length standards at all though at the exit Echte[3] of the highway Hannover Kassel a sign proudly announces "Echte 1000 m".

Distance, which exceeds a few meters, can be measured only with poor accuracy by placing metersticks one after the other. One measures length optically with light.

Michelson's measurements prove that, independent of the velocity of the observer, measuring rods and rigid bodies measure distances which agree, within the precision limited by the inaccurate rods, with the ones from devices which measure the runtime of light.

Since c, the speed of light in the vacuum, is constant and because of the imperfections of rigid bodies as length standards, one measures spatial distance by half the time which it takes light to run back and forth (1.5). This distance coincides with the distances determined at the best with measuring rods. Since 1983 physicists define 299 792 458 m to be the distance, which light in the vacuum propagates in 1 s.

We choose to specify distance simply by the time of flight of light. Then a light year is just a year and a light second just a second and

$$1 \text{ s} = 299\,792\,458 \text{ m} . \tag{1.2}$$

By this choice of the unit of length velocities are pure numbers, namely their ratio as compared to the velocity of light, and c has the natural value 1. Meter per second is a numerical factor such as kilo or milli and denotes approximately 3.3 nano,

$$\frac{\text{m}}{\text{s}} = \frac{1}{299\,792\,458} \approx 3.3356 \cdot 10^{-9} , \quad c = 299\,792\,458 \, \frac{\text{m}}{\text{s}} = 1 . \tag{1.3}$$

---

[3] The German word "Echte" means genuine, real.

In units with $c = 1$ the equations of relativistic physics are particularly simple and reveal common features of position and time. Only in some results we insert the factors c as they occur ordinarily if we do not convert seconds into meters.

That distances can be specified as times is child's play. In Grimm's Little Red Ridinghood grandmother's house is half an hour away from the village. To speak of a second rather than a light second is no more a revolutionary alteration or unlawful frivolity than to denote the distance of half an hour's walk just by half an hour.

That length and time are different does not prevent to measure them in common units. After all, in aviation height and distance are different and are measured in feet and in nautical miles. Nevertheless, the slope of a flight path does not depend on units because foot per nautical mile is nothing but the numerical factor $1.646 \cdot 10^{-4}$ which only differs in size from meter per second which equals $3.3356 \cdot 10^{-9}$.

Though distances are measured by definition by the time of flight of light and though consequently its velocity has the constant value $c = 1$, one can nevertheless check experimentally the consistency of this definition. First choose four points O, X, Y and Z, which do not lie in a plane and determine their distances. The results determine the scales and angles in a reference system and are not restricted to satisfy any equation. If then one measures the distances of a fifth point A to the four reference points, then the resulting distance to O restricts A to a sphere around O, the distance to X restricts A to the intersection of two spheres, a circle, and the distance to Y to the intersection of this circle with a third sphere. Therefore the measured distance to Z has to coincide with one of two possible values if the definition is consistent, that the velocity of light is constant everywhere and at all times. The measured $n(n-1)/2$ distances of $n \geq 5$ particles have to satisfy $n(n-1)/2 - 3n + 6$ relations if the velocity of light is constant. This is not guaranteed by definition but abstracted without any conflicting evidence from experience.

Equilocal and Equitemporal

In the course of time a traveller in a train traverses a set of events, his worldline. Related to a point in the train, all events on his worldline occur at the same place. For an observer at the railway embankment the traveller passes different places. Equilocality, the property of two different events to occur at the same place, depends on the observer.

Figure 1.4 shows the worldlines of a uniformly moving clock $\mathscr{C}$ and of an observer $\mathscr{O}$ at rest with respect to the clock who also carries a clock. We denote the events on the worldlines by the time on the clock, which is carried along. At time t_- the observer emits light which is reflected and received at time t_+. With this light the observer sees the clock $\mathscr{C}$ show the time t.

Because the clock $\mathscr{C}$ does not move relative to the observer $\mathscr{O}$, i.e. because it is comoving, the time $t_+ - t_-$, which it takes light to run back and forth, and the direction of the light do not depend on t_- and $\mathscr{C}$ passes a worldline parallel to the observer's. In particular, the ends of a measuring rod also pass parallel worldlines.

Fig. 1.4 Equilocal

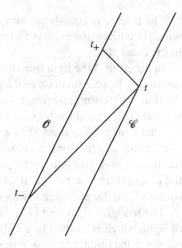

Fig. 1.5 Equal clocks at
relative rest

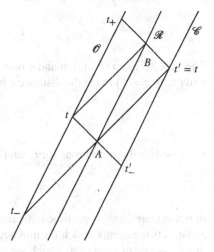

If the worldline $\mathcal{O}$ is parallel to $\mathcal{C}$ and if $\mathcal{C}$ is parallel to the worldline of another observer $\mathcal{R}$, then $\mathcal{R}$ is parallel also to $\mathcal{O}$. Therefore, if and only if two observers do not move with respect to each other, they agree whether two different events occur at the same place.

In the same way observers agree, if and only if they rest with respect to each other, whether different events occur at the same time. To occur simultaneously or to be equitemporal depends on the observer and is no intrinsic property of pairs of events, because no measurable universal time exists.

Figure 1.5 clarifies which events are simultaneous or equitemporal for an observer. It shows the worldlines of two observers $\mathcal{O}$ and $\mathcal{C}$ and a referee $\mathcal{R}$ in their middle.

The referee is equally far away from $\mathcal{O}$ and $\mathcal{C}$ because he sees flashes of light which he emits in the event A and which are reflected by $\mathcal{O}$ and $\mathcal{C}$ return both together in the event B.

In the event A the light from $\mathcal{O}$ and $\mathcal{C}$ shows the referee $\mathcal{R}$ the times t_- and t'_- on the clocks; he sees them show the times t and t' in event B. Both clocks are equal if the time differences coincide, $t - t_- = t' - t'_-$. For simplicity, we assume the clocks set such, that they show him the same time, $t = t'$.

That it takes light from $\mathcal{O}$ to $\mathcal{C}$ the same time as for the way back is not a mere convention, as occasionally claimed, but an observation. In event A the referee sees the light start at equal times $t_- = t'_-$ and in event B he sees the light arrive while the clocks show $t = t'$. That it takes light the same time to run back as to run forth follows from the homogeneity and isotropy of spacetime.

The triangles $t_- A t$ and $t B t_+$ are congruent, so $t - t_- = t_+ - t$. The reflection of light, which is emitted by an observer at t_- and received after the reflection at t_+ occurs in the middle of these times at

$$t = \frac{t_+ + t_-}{2} . \qquad (1.4)$$

The time between reception and emission of light to and from E is by definition (in units with $c = 1$) twice the distance r between E and the observer,

$$r = \frac{t_+ - t_-}{2} . \qquad (1.5)$$

If one solves these relations for t_+ and t_- one finds the denominations justified,

$$t_+ = t + r \quad , \quad t_- = t - r . \qquad (1.6)$$

In two-dimensional spacetime diagrams with given worldline of an observer $\mathcal{O}$ one constructs the events which for him occur simultaneous to an event E as a diagonal in a rectangle of light rays, which we call lightangle [19].

One draws the light rays through E to their intersection t_+ and t_- with the worldline $\mathcal{O}$. The light rays emanating from t_- and the incoming light rays of t_+ form the lightangle $t_- E t_+ E'$. For the observer $\mathcal{O}$ the events E and E' occur at the same time and the same distance in opposite direction, because both events correspond to the emission time t_- and reception time t_+. If one enlarges or shrinks the lightangle, holding the intersection of the diagonals fixed, one confirms that all events on the straight line through E and E' are simultaneous for $\mathcal{O}$.

The worldline of the observer and events which for him occur at the same moment constitute the diagonals of a lightangle, one diagonal consists of equilocal events, the other of equitemporal ones.

Equilocality and simultaneity is not a geometric property which pertains to pairs of events. It also depends on the observer. The worldlines of mutually moving observers are not parallel and the diagonals in their corresponding lightangles are not parallel.

Fig. 1.6 Equilocal and
equitemporal diagonals

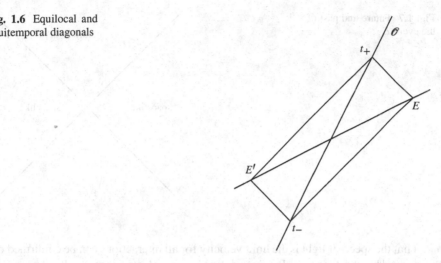

Therefore mutually moving observers do not agree which events are equilocal nor
which are equitemporal.

1.3 Limit Speed

By the principle of relativity, that in the vacuum one cannot distinguish rest from
uniform motion, all causes, not only electromagnetic ones, produce effects at most
with the speed of light. Otherwise, if by some interaction an event E' in the vacuum could cause an effect with a limit speed faster than light and influenced the
event E as in Fig. 1.6, then there would be an observer $\mathcal{O}$ for whom E' and E occur
simultaneously. In this way the superluminal velocity and light would single out a
particular observer who one could distinguish by physical observation in the vacuum
from other observers.

In a medium, a fluid say, the velocity of sound could surpass the velocity of
light in the vacuum. Though one never observed such a miraculous medium, it's
existence would not lead to logical contradictions, even if then one could construct
a sequence of events on a closed loop, where each event causes the next. In such a
situation one could—as is argued—shoot one's own grandfather, a logical contradiction, because the grandfather is necessary for the existence of his killer. The
contradiction rests on the assumption, that on closed causal loops one still has the
freedom to choose what to do. However, as the contradiction shows, this assumption
is wrong. If there are closed causal loops, then one can change the conditions only
in such a way, that along the closed causal loop one arrives at the initial situation.
On causal loops one is bound to the wheel of perpetual rebirth.

Fig. 1.7 Future and past of
the event E

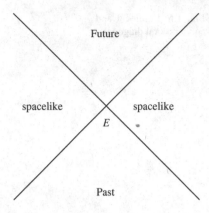

That the speed of light is the limit velocity for all interactions can be confirmed or
disproved by observations. Neutrinos do not interact electromagnetically. As long as
one did not know one assumed that they propagated with the limit velocity of light.
Today, we know it. The limit velocity of neutrinos coincides with the speed of light
at least in the first nine decimals [31] because in the supernova SN 1987a in Feb-
ruary 1987 one has detected neutrinos and light which arrived simultaneously after
160 000 years of flight.

If we receive a message, then the information is caused by the sender. It cannot
be transmitted faster than light.

The events E', which can be influenced by E constitute the future of E and occur
later than E by at least the runtime of light (Fig. 1.7).

The past of E consists of all events E' which could have influenced E. They have
occurred earlier than E by at least the runtime of light from E' to E.

We call events E' spacelike with respect to E if they occur so early, that they
cannot be influenced by E and which are not so long ago that one could know about
them at E. If E' is spacelike to E then, vice versa, E is spacelike to E'. Mutually
spacelike events cannot influence one another.

Different from simultaneity, the property of pairs of events to be spacelike does
not depend on an observer. But this property is not transitive: if A is spacelike to B
and B spacelike to C, then this does not make A spacelike to C.

The future and the past of E are separated by the forward and backward lightcone
of E from the events which are spacelike to E.

Tachyon and Rigid Bodies

Since a particle is at a position because it was previously at a position nearby and
because effects propagate with the speed of light at most, therefore the worldline of

each particle and of each observer always runs from the past to the future and lies within the lightcone. There is nothing faster than light. Nothing outruns light.

Tachyon is the name of a hypothetical particle, which moves faster than light. It would traverse a worldline such as $E'E$ in Fig. 1.6. If a tachyon could scatter light it would be at first invisible to an observer $\mathcal{O}$ whose worldline it intersects, since its worldline does not intersect the backward lightcone of early events on the worldline of the observer $\mathcal{O}$. At later times each backward lightcone of the worldline of $\mathcal{O}$ intersects the worldline of the tachyon in two events. These two intersections depart in opposite directions in the course of time when $\mathcal{O}$ traverses his worldline. Hence, the tachyon would appear to the observer as a pair of particles which emerge out of nowhere at some point and run away in opposite directions. There is not a single serious observation which would suggest the existence of tachyons.

One can easily cause effects on a tachyonic worldline, for instance if one connects lamps at a runway of an airport with separate cables of equal length and if one simultaneously switches them on. An observer on the runway then sees at first the nearest lamp light up, then the two neighboring lamps, as if a signal would propagate from the nearest lamp to both sides with superluminal speed. However, no lamp lights up be*cause* the neighboring lamp has turned on: walls which one puts between the lamps do not interrupt a signal from lamp to lamp. The cause is the flight controller who has switched on the light. This cause produces effects at most with the speed of light.

Since causes act at most with the speed of light, there is no body, which is precisely rigid. For instance, a blow on one end of a bar does not influence the other end at first. The compression caused by the blow propagates in the bar as a wave with the speed of sound, and only after it has traversed the body and vibrations have faded away does the bar recover its original state. One cannot evade the conclusion that there is no ideally rigid body by imagining a very hard body which resists its deformation with strong internal forces. Hard bodies have high sonic speeds, but the speed of sound is always less than c, the speed of light in the vacuum.

1.4 Quantum Teleportation and Bell's Inequality

The revolutionary conclusion of quantum physics is that even if one prepares particles as best as possible there are always measurements with results which no one can predict in each single case but where one knows only the probability of the different possible results. In the particular example, where one measures the polarization of pairs of photons, one can disprove the assumption that the inability to predict the individual results relies only on incomplete knowledge. One is forced to conclude that measuring devices do not show results which were determined previously but ascertain results which were uncertain before.

If light passes a polarization filter, it becomes polarized. It passes unchanged a second filter, which polarizes in the same direction a and its intensity decreases by a factor

$$p_a(b) = \cos^2 \beta \,, \tag{1.7}$$

if one rotates the second filter in the plane which is orthogonal to the light by an angle β into the direction b. No light passes through crossed filters, if $b = a_\perp$ is orthogonal to a. In the following we denote a filter which polarizes in direction a as filter a for short.

Surprisingly, light with low intensity exhibits properties of particles. The photo-electric effect does not become smaller with decreasing intensity of light but rarer. One therefore has to interpret the intensity of light as proportional to the probability to find a photon and the reduction factor $p_a(b)$ as probability for the photon, which has been polarized in direction a, to pass the filter b.

With the remaining probability $1 - \cos^2 \beta = \sin^2 \beta$ the photon is absorbed. This is the same as the probability to pass the crossed filter $b_\perp$,

$$p_a(b_\perp) = 1 - p_a(b) \,. \tag{1.8}$$

In a suitably chosen transition of exited calcium atoms pairs of photons are created, which are emitted in opposite directions with such a polarization that the first photon passes a filter a and the second a filter b with probability [2]

$$p(a, b) = \frac{1}{2} \cos^2 \beta \,, \tag{1.9}$$

where β is the angle between the directions a and b.

The probability for the result that the first photon passes the filter and the second photon is absorbed is the same as the probability $p(a, b_\perp)$ that the first photon passes the filter a and the second passes the crossed filter $b_\perp$. In the same way $p(a_\perp, b)$ and $p(a_\perp, b_\perp)$ are the probabilities that the first photon is absorbed while the second passes and for the case that both photons are absorbed,

$$p(a, b_\perp) = \frac{1}{2} \sin^2 \beta \,, \quad p(a_\perp, b) = \frac{1}{2} \sin^2 \beta \,, \quad p(a_\perp, b_\perp) = \frac{1}{2} \cos^2 \beta \,. \tag{1.10}$$

These probabilities shatter, as we shall see, the physicists' view of the world. They exclude the interpretation that each photon is equipped with a property which deter-mines in each case and for all filters whether the photon passes or is absorbed. This result is really accidental, as the following considerations show.

Combining the two possibilities, that the second photon passes or not, we obtain the probability

$$p_1(a) = p(a, b) + p(a, b_\perp) = \frac{1}{2} \tag{1.11}$$

that the first photon passes the filter a, irrespective of what happens to the other photon. This probability is the same as the probability $p_1(a_\perp)$ for the first photon to be absorbed. The probability is independent of the direction a. Also the second photon becomes absorbed with the same probability $p_2(b_\perp) = p_2(b) = 1/2$ with which it passes filter b. Both photons of the pair are unpolarized, as far as measurements of one photon are concerned.

If one considers the subset of cases, in which the first photon passes the filter a, then the second photon passes filter b with the conditional probability

$$\frac{p(a, b)}{p_1(a)} = \cos^2 \beta . \tag{1.12}$$

This is the same probability as for photons which are polarized by a filter a (1.7). If the first photon passes the filter a, then in this subset of events the second photons is polarized in direction a. In particular, it passes a polarization filter a with certainty.

For this behavior there is the way of speaking that the polarization measurement of one photon changed the state of the other photon of the pair instantaneously, no matter how far away it may be, to the state of the same polarization and that the state of the pair collapsed or became reduced. The result of the first measurement, so the alleged interpretation, was transmitted to the second photon or, even more impressive, was quantum teletransported. The reduction of state happened instantaneously faster than light.

Untouched by these sensational claims one can conclude as a matter of fact that a measurement of one photon does not cause anything at the second photon. No measurement, performed at one photon, can detect whether the other photon had been measured, is being measured or will be measured, leave alone in which direction and with which result.

That the second photon is polarized in direction a in case that the first photon passes the filter a, can be confirmed only after one knows at the second filter, whether the first photon has passed and what the direction of the filter was. This information can be transmitted with the speed of light at the fastest.

The correlation of the results of the polarization measurements is caused by the joint preparation of the two photons as a pair in one atomic transition. This preparation works only if both photons are created at the same place. They move with the speed of light. Therefore, this preparation does not produce effects which propagate faster than light.

If one throws a coin repeatedly and sends the picture of the upper side to one recipient and the picture of the down side to a second recipient, then each of them obtains pictures of head and tail with equal probability. Each of them knows in the instant, when he opens his letter, the picture which the other receives. The knowledge of the result collapses the probability to a conditional probability, in this example to certainty.

In the same way the reduction of state by the result of a measurement replaces the previous state by the conditional state which pertains to the conditional probability in the case of the measured result.

Before opening the letter, the receiver is uncertain about its content, but the content is not uncertain but only unknown. The content is certain, whether one opens the letter or not. The probability (1.9) however, contradicts the assumption, that the results of the polarization measurements are determined for all directions and in each case and that one does not know the result prior to the measurement only because the individual causes are insufficiently known.

Such an assumption seems to be irrefutable, but it leads to a mathematical restriction, an inequality, which may or may not be confirmed by experiment.

The measured results violate the inequality and disprove the assumption.

To evaluate the assumption, we consider repeated measurements and enumerate them by i, $i = 1, 2, \ldots, N$. We assume that in the measurement number i the result of the polarization measurement of the first photon in direction a is determined by some causes, even if we do not know them, and we attribute $a_{1i} = 1$ to the case that the first photon passes, if not $a_{1i} = -1$. With b_{1i} we denote the results which in experiment number i would be obtained if the polarization of the first photon was measured in direction b. In the same way c_{2i} and d_{2i} denote the results if we measure the polarization of the second photon in experiment number i with a filter c or d.

In all cases the results a_{1i}, b_{2i}, c_{2i} and d_{2i} can only take the values 1 or -1 and if $a_{1i}(c_{2i} + d_{2i})$ is ± 2 then $b_{1i}(c_{2i} - d_{2i})$ vanishes and vice versa. Therefore their sum never exceeds 2 [9],

$$a_{1i} c_{2i} + a_{1i} d_{2i} + b_{1i} c_{2i} - b_{1i} c_{2i} \leq 2 . \tag{1.13}$$

The average $\langle a_1 c_2 \rangle$ of the products $a_{1i} c_{2i}$ of the results in N experiments is the sum over the individual products divided by N,

$$\langle a_1 c_2 \rangle = \frac{1}{N} \sum_{i=1}^{N} a_{1i} c_{2i} . \tag{1.14}$$

Correspondingly we obtain the averages of $\langle a_1 d_2 \rangle$, $\langle b_1 c_2 \rangle$ and $\langle b_1 d_2 \rangle$. If we sum the inequality (1.13) and divide by N, we obtain a Bell inequality [4] for the average of products of results of polarizations measurements

$$\langle a_1 c_2 \rangle + \langle a_1 d_2 \rangle + \langle b_1 c_2 \rangle - \langle b_1 d_2 \rangle \leq 2 . \tag{1.15}$$

We can evaluate the average of $a_{1i} c_{2i}$ also by counting the frequency N_+ and N_- with which each possible value $+1$ or -1 occurs. If we multiply the frequency with the corresponding value, we get $N_+ - N_- = \sum_{i=1}^{N} a_{1i} c_{2i}$ and $\langle a_1 c_2 \rangle = (N_+ - N_-)/N$.

If N is sufficiently large, then the relative frequency N_+/N is the probability for the cases where $a_{1i} c_{2i}$ has value $+1$ and N_-/N is the probability for the value -1. The probability for $a_{1i} c_{2i}$ to have the value $+1$ is $p(a, c) + p(a_\perp, c_\perp)$, with probability $p(a, c_\perp) + p(a_\perp, c)$ the product has the value -1. Therefore the experimental probability (1.9), which is also predicted by quantum theory, yields the average

$$\langle a_1 c_2 \rangle = \cos^2 \gamma - \sin^2 \gamma = \cos(2\gamma) . \tag{1.16}$$

It is given by the scalar product of unit vectors $\mathbf{A}$ and $\mathbf{C}$ which include double the angle as a and c, $\cos(2\gamma) = \mathbf{A} \cdot \mathbf{C}$.

Fig. 1.8 Directions of polarization

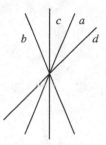

By the same reason $\langle a_1 d_2 \rangle = \mathbf{A} \cdot \mathbf{D}$, $\langle b_1 c_2 \rangle = \mathbf{B} \cdot \mathbf{C}$ and $\langle b_1 d_2 \rangle = \mathbf{B} \cdot \mathbf{D}$ are scalar products of unit vectors with doubled angles.

If one varies the vectors $\mathbf{A}$ and $\mathbf{B}$, then the sum

$$\langle a_1 c_2 \rangle + \langle a_1 d_2 \rangle + \langle b_1 c_2 \rangle - \langle b_1 d_2 \rangle = \mathbf{A} \cdot \mathbf{C} + \mathbf{A} \cdot \mathbf{D} + \mathbf{B} \cdot \mathbf{C} - \mathbf{B} \cdot \mathbf{D} \qquad (1.17)$$

becomes maximal, if $\mathbf{A}$ has the direction of $\mathbf{C} + \mathbf{D}$ and $\mathbf{B}$ the direction of $\mathbf{C} - \mathbf{D}$.

Then the sum has the value $|\mathbf{C} + \mathbf{D}| + |\mathbf{C} - \mathbf{D}|$ which takes its maximal value $2\sqrt{2}$ if the unit vectors $\mathbf{C}$ and $\mathbf{D}$ are orthogonal. Therefore, if one takes c to include the angle $\pi/8 = 22.5°$ with b, a an angle of $\pi/4 = 45°$ and d an angle of $3\pi/8 = 67.5°$ then the sum of the averages becomes maximal and has the value $2\sqrt{2}$. This value is confirmed by the measurements and disproves Bell's inequality (1.15).

To derive Bell's inequality we only assumed that the results a_{1i}, b_{1i}, c_{2i} and d_{2i} of each measurement i are certain and do not depend on the direction in which one chooses to measure the other photon. In real life [2] one can measure in each single case only in one direction, either in direction a or b and either in direction c or d. One has to determine a_{1i} and b_{1j} or c_{2i} and d_{2j} in different measurements $i \neq j$. A random generator chooses the directions of the measurements after the photons have left the source and so late, that the choice cannot be known by signals with the speed of light at the time of measurement at the position of the other photon (Fig. 1.8).

That the measured values of the product polarization do not satisfy Bell's inequality shatters the views which physicists have of the world. Measurements do not read off properties, which determine the result, because then the results would have to satisfy Bell's inequality. Measurements ascertain results which previously were not certain.

The measured violation of Bell's inequality refutes the elusion from reality that each result was certain in each case but only the cause of each result was insufficiently known.

In quantum physics there is no cause for each single result but only causes for probabilities of results of measurements.

Chapter 2
Time and Distance

Abstract An elementary geometric fact, stated as the intercept theorem, makes an observed clock run visibly slower, if it moves away in the line of sight and to run visibly faster by the inverse factor, if it approaches the observer with the same velocity. This Doppler effect of light in the vacuum is particularly simple, because, different from the Doppler effect of sound, it depends only on the relative velocity of the light source and its observer. We employ a referee to determine whether moving clocks are equal and how the times between pairs of events compare. This time endows spacetime with a geometric structure, the distance, which is similar to but also different from Euclidean distance. From the Doppler effect we determine the addition of velocities, time dilation and length contraction and clarify the related paradoxes.

2.1 Theorem of Minkowski

Consider as in Fig. 2.1 a clock $\mathscr{C}$ and an observer $\mathscr{O}$ with another clock. Both move uniformly along straight worldlines and meet in the event O, the origin, where their worldlines intersect. There both clocks are set to zero for simplicity, so that in the following we can speak of times rather than of time differences. When the observer $\mathscr{O}$ looks at the clock $\mathscr{C}$, which moves away from him uniformly in the line of sight, then he reads off the time $t_{\mathscr{C}}$ which passed on $\mathscr{C}$ until the emission of the light. At the moment of observation his own clock shows a time $t_{\mathscr{O}}$. This time of reception is proportional to the time of emission[1]

$$t_{\mathscr{O}} = \kappa(\mathscr{O}, \mathscr{C})\, t_{\mathscr{C}} \quad \text{for } t_{\mathscr{C}} > 0 , \tag{2.1}$$

---

[1] κ and ν are the Greek letters kappa and nu.

N. Dragon, *The Geometry of Special Relativity—a Concise Course*,
SpringerBriefs in Physics, DOI: 10.1007/978-3-642-28329-1_2,
© The Author(s) 2012

Fig. 2.1 Intercept theorem

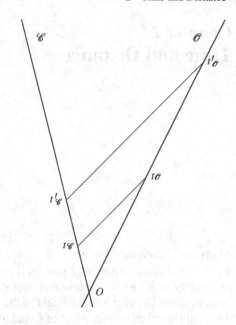

with a coefficient $\kappa(\mathcal{O},\mathcal{C})$, which does not depend on $t_\mathcal{C}$ [7]: if the observer later reads off the time $t'_\mathcal{C}$, then the triangle $Ot'_\mathcal{C}t'_\mathcal{O}$ is similar to $Ot_\mathcal{C}t_\mathcal{O}$ and all distances are enlarged by the same factor. Therefore the ratios $t_\mathcal{O}/t_\mathcal{C}$ and $t'_\mathcal{O}/t'_\mathcal{C}$ coincide.

If a quartz is carried along with the clock $\mathcal{C}$ and oscillates n times during the time $t_\mathcal{C}$ with a frequency $\nu_\mathcal{C} = n/t_\mathcal{C}$, then the observer $\mathcal{O}$ sees these oscillations while on his own clock the time $t_\mathcal{O}$ elapses. So he observes the frequency

$$\nu_\mathcal{O} = \frac{1}{\kappa(\mathcal{O},\mathcal{C})}\nu_\mathcal{C} . \tag{2.2}$$

This visible change of frequency of the clock which moves in the line of sight is the longitudinal Doppler effect. It is related to the Doppler effect of sound, which one can hear as whining drop of the pitch of passing police cars or racing cars.

As one cannot distinguish rest from uniform motion, the Doppler factor $\kappa(\mathcal{O},\mathcal{C})$ only depends on the relative velocity of $\mathcal{C}$ and $\mathcal{O}$ and, contrary to the Doppler effect of sound, not on the velocity with respect to a medium.

Moreover, κ depends on whether both clocks run equally fast. For two clocks at rest this can be easily seen. For the moving clocks $\mathcal{O}$ and $\mathcal{C}$ this is more difficult. One has to correct for the various and changing times which it takes light to run from the clock to the observer who compares both clocks.

However, no correction is necessary for a referee $\mathcal{R}$ as in Fig. 2.2 who is always in the middle of the clocks. Flashes of light which he emits at some time to $\mathcal{O}$ and $\mathcal{C}$ are reflected and return in the same instant. Because the referee is always in the middle, the runtimes of light to and from $\mathcal{O}$ and $\mathcal{C}$ are equal.

Fig. 2.2 Comparison of
clocks

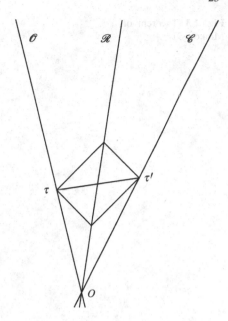

Both clocks run the same if they show the referee equal times,

$$\tau' = \tau .\tag{2.3}$$

This is the geometric definition of equal lengths on intersecting straight world-lines of moving observers. Without exception the definition agrees with the physical behavior of real equal clocks.

We continue the worldlines of the light rays, which are received and reflected by $\mathscr{C}$ as it shows the time τ', to the worldline of the observer $\mathscr{O}$ and denote in Fig. 2.3 with t_- and t_+ the times shown by the clock of $\mathscr{O}$ as he emits the light ray to $\mathscr{C}$ and receives it, respectively. Due to (2.1) the clock $\mathscr{O}$ shows the time

$$\tau = \kappa(\mathscr{O},\mathscr{R})\kappa(\mathscr{R},\mathscr{O})t_-\tag{2.4}$$

when the light ray emitted at time t_- and reflected by $\mathscr{R}$ arrives. This is because τ is a multiple of the time at which the light ray was reflected by $\mathscr{R}$, and this time is a multiple of the time t_- at which the light ray was emitted by $\mathscr{O}$. By the same reason

$$t_+ = \kappa(\mathscr{O},\mathscr{R})\kappa(\mathscr{R},\mathscr{O})\tau .\tag{2.5}$$

Thus, $t_+/\tau = \tau/t_-$ and $\tau^2 = t_+ t_- = t^2 - r^2$ (1.6). Moreover, the equal clocks show equal times, $\tau' = \tau$. This proves the

Theorem of Minkowski: *Let two observers $\mathscr{O}$ and $\mathscr{C}$ move linearly and uniformly and meet in some event O, when they set their equal clocks to zero. Then the time τ,*

Fig. 2.3 Theorem of
Minkowski

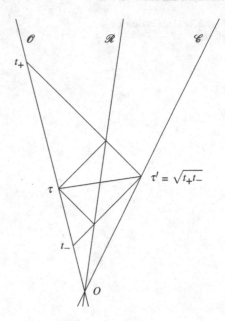

which elapses on the clock $\mathscr{C}$ until an event E, is the geometric mean of the time t_-
shown by the clock of the observer $\mathscr{O}$ when he emits light to E and the time t_+ which
his clock shows when he receives light from E,

$$\tau^2 = t_+ t_- = t^2 - r^2 . \tag{2.6}$$

This relation is as important for the geometry of spacetime as the Pythagorean
theorem $c^2 = a^2 + b^2$ for Euclidean geometry. According to the Pythagorean theorem
in Euclidean geometry all points on a circle are equally far away from the center. The
equation $\tau^2 = t^2 - r^2$ implies that in spacetime points of equal temporal distance to
the origin O lie on hyperbolas.

Three Equal Clocks

The definition, that equal clocks show their referee equal times differences, is con-
sistent: the clock $\mathscr{O}_3$ equals the clock $\mathscr{O}_1$ if it equals the clock $\mathscr{O}_2$ and if the clock $\mathscr{O}_2$
equals the clock $\mathscr{O}_1$ (Fig. 2.4).

If the clocks move in the same direction and meet in a common event, then the
relation

$$t^4 = t_+^2 t_-^2 = t_{++} t_{+-} t_{-+} t_{--} \tag{2.7}$$

holds. As in (2.4) one has

Fig. 2.4 Three equal clocks

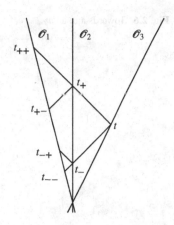

Fig. 2.5 Geometric mean

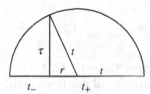

$$t_{-+} = \kappa(\mathcal{O}_1, \mathcal{O}_2)\kappa(\mathcal{O}_2, \mathcal{O}_1)t_{--} \tag{2.8}$$

and

$$t_{++} = \kappa(\mathcal{O}_1, \mathcal{O}_2)\kappa(\mathcal{O}_2, \mathcal{O}_1)t_{+-} . \tag{2.9}$$

Therefore $t^4 = t_{++}^2 t_{--}^2$ and the clock of $\mathcal{O}_3$ equals the clock of $\mathcal{O}_1$. If two clocks equal a third then they equal each other. This holds also, if the worldlines of the clocks do not lie in a plane or do not intersect because the worldlines can be translated and rotated without changing the clocks.

Construction of the Referee

To construct the worldline of the referee between two observers $\mathcal{O}$ and $\mathcal{C}$ one draws the light rays through a point τ' on one worldline (Fig. 2.1). They intersect the other worldline in t_+ and t_- and determine the geometric mean τ of t_+ and t_-. The worldline of the referee is the straight line through the intersection of the light rays through τ and τ'. This worldline also passes the origin O, because τ is the geometric mean of t_+ and t_-.

The geometric mean $\sqrt{t_+ t_-}$ is constructed in Euclidean geometry by help of a circle with a diameter, which consists of the line segments t_+ and t_- (Fig. 2.5). Its radius is the arithmetic mean $t = (t_+ + t_-)/2$, the line segment t_+ is longer by

Fig. 2.6 Towards and away

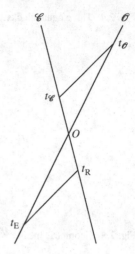

$r = (t_+ - t_-)/2$, $t_+ = t + r$, the line segment t_- is shorter $t_- = t - r$. The orthogonal line through the endpoint of t_+ cuts the circle with a segment of length $\tau = \sqrt{t^2 - r^2} = \sqrt{t_+ t_-}$.

2.2 Addition of Velocities

In Fig. 2.3 the Doppler factor $\tau'/t_- = \kappa(\mathscr{C}, \mathscr{O})$ is the ratio of the time of reception to the time of emission (2.1) of light rays sent from the observer $\mathscr{O}$ to $\mathscr{C}$, and $t_+/\tau' = \kappa(\mathscr{O}, \mathscr{C})$ is the ratio for the way back. Both clocks are equal, $\tau = \tau'$. Therefore (2.4) and (2.5) state that the Doppler factor $\kappa(\mathscr{O}, \mathscr{C})$, by which $\mathscr{O}$ sees frequencies of $\mathscr{C}$ shifted, equals the Doppler factor $\kappa(\mathscr{C}, \mathscr{O})$, by which $\mathscr{C}$ perceives shifted frequencies of $\mathscr{O}$

$$\kappa(\mathscr{C}, \mathscr{O}) = \kappa(\mathscr{O}, \mathscr{C}) . \tag{2.10}$$

On motion in the line of sight the Doppler shift is reciprocal.

From this reciprocity alone, from $t_+ = \kappa\tau$ together with $\tau = \kappa t_-$, one can conclude Minkowski's theorem and the dependence of the Doppler factor on the relative velocity,

$$\kappa^2 = \frac{t_+}{t_-}, \quad \tau^2 = t_+ t_- . \tag{2.11}$$

The relations $t_+ = t + r$ and $t_- = t - r$ (1.6) imply

$$\kappa^2 = \frac{t+r}{t-r} = \frac{1+r/t}{1-r/t} , \quad \tau^2 = t^2 - r^2 = \left(1 - \frac{r^2}{t^2}\right) t^2 , \tag{2.12}$$

Fig. 2.7 Addition of velocities

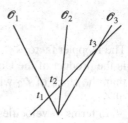

and, because $v = r/t$ is the velocity with which the clock $\mathscr{C}$ moves away from the observer $\mathscr{O}$,

$$\kappa(v) = \sqrt{\frac{1+v}{1-v}} = \frac{1+v}{\sqrt{1-v^2}}, \qquad (2.13)$$

$$v = \frac{\kappa^2 - 1}{\kappa^2 + 1}, \qquad (2.14)$$

$$\tau = \sqrt{1-v^2}\, t. \qquad (2.15)$$

If the observer $\mathscr{O}$ emits a pulse of light at a time $t_E < 0$, while the clock moves towards him (recedes with negative velocity) then, as Fig. 2.6 shows, the ratio $\kappa(-v) = t_R/t_E$ of the times of reception and emission is the inverse of the ratios of the times which the clocks show later, when they move away from each other

$$\kappa(-v) = \frac{t_R}{t_E} = \frac{t_\mathscr{C}}{t_\mathscr{O}} = \frac{1}{\kappa(v)}. \qquad (2.16)$$

A clock, which moves away from an observer, appears slower, because it shows him the time $t_\mathscr{C} = t_\mathscr{O}/\kappa$, when his own and equal clock shows $t_\mathscr{O}$ and $\kappa(v)$ is larger than 1 for positive velocity $v > 0$.

On motion in the line of sight an approaching clock appears faster, because during the approach the Doppler factor is inverse to the Doppler factor during recession.

With (2.14) one can determine the velocity v (as is routinely done by traffic authorities) by measuring the Doppler shift κ. It retains its value, if one exchanges observer and observed object. Therefore, two observers who move in the line of sight measure the same relative velocity. We use (2.13) to determine the relative velocities of several observers (Fig. 2.7).

If three observers, $\mathscr{O}_1$, $\mathscr{O}_2$ and $\mathscr{O}_3$, move in the same direction and register the times on their clocks at which a light pulse passes then these times are proportional,

$$t_2 = \kappa_{21} t_1, \quad t_3 = \kappa_{32} t_2, \quad t_3 = \kappa_{31} t_1. \qquad (2.17)$$

From $\kappa_{31} t_1 = \kappa_{32} \kappa_{21} t_1$ one immediately concludes

$$\kappa_{31} = \kappa_{32}\kappa_{21} \; . \tag{2.18}$$

The Doppler factor κ_{31}, by which $\mathscr{O}_3$ sees his clock run faster than the clock of $\mathscr{O}_1$, is the product of the Doppler factor κ_{32}, by which $\mathscr{O}_3$ observes his clock run faster than the clock of $\mathscr{O}_2$ with the Doppler factor κ_{21} for the observer $\mathscr{O}_2$ and the clock of $\mathscr{O}_1$.

In terms of velocities (2.13) and squared this means (in our units with $c = 1$)

$$\frac{1 + v_{31}}{1 - v_{31}} = \frac{1 + v_{32}}{1 - v_{32}} \frac{1 + v_{21}}{1 - v_{21}} \tag{2.19}$$

or, solved for v_{31},

$$v_{31} = \frac{v_{32} + v_{21}}{1 + v_{32}v_{21}} \; . \tag{2.20}$$

The velocity v_{31}, with which $\mathscr{O}_3$ sees the observer $\mathscr{O}_1$ recede, is not the sum $v_{32} + v_{21}$ of the velocity v_{32}, with which $\mathscr{O}_3$ observes $\mathscr{O}_2$ recede, and the velocity v_{21}, with which $\mathscr{O}_2$ perceives the recession of $\mathscr{O}_1$. The naive addition of velocities is only approximately correct as long as in ordinary life v_{32} and v_{21} are small compared to the speed of light, $c = 1$.

Up to the sign in the denominator velocities add like inclinations. If the bed of a tipper lorry is inclined by an angle α, then on even ground it has the slope $m_1 = \tan \alpha$. If the truck drives a street with slope $m_2 = \tan \beta$ then its bed has an overall angle $\alpha + \beta$ to the horizontal and the overall inclination

$$m_3 = \frac{\sin(\alpha + \beta)}{\cos(\alpha + \beta)} = \frac{\cos \alpha \sin \beta + \sin \alpha \cos \beta}{\cos \alpha \cos \beta - \sin \alpha \sin \beta} = \frac{\tan \alpha + \tan \beta}{1 - \tan \alpha \tan \beta} = \frac{m_1 + m_2}{1 - m_1 m_2} \; . \tag{2.21}$$

We define the rapidity σ as the logarithm of the Doppler factor κ,

$$\sigma = \ln \kappa = \frac{1}{2} \ln \frac{1 + v}{1 - v}, \quad v = \frac{e^\sigma - e^{-\sigma}}{e^\sigma + e^{-\sigma}} = \tanh \sigma \; . \tag{2.22}$$

To the addition of rapidities there corresponds the multiplication of the Doppler factors, $\kappa = e^\sigma$. These rapidities, not the velocities, add on motion of several observers in the same direction.

2.3 Time Dilation

If the time t elapses on a clock between two events O and E, then on a second equal clock, which moves relative to the first one with a velocity v, the shorter time (2.15)

$$\tau = \sqrt{1 - v^2} \, t \tag{2.23}$$

Fig. 2.8 Reciprocal dilation
of time

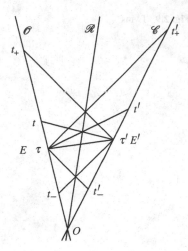

goes by between the corresponding events which are simultaneous to O and E for
the first clock.

Time dilation is reciprocal. This can be deduced from Fig. 2.8 where we have
continued the light rays of Fig. 2.3 to the worldlines of the observers $\mathcal{O}$ and $\mathcal{C}$.

The events are denoted by the times on the clocks which the observers carry along.

All the time the referee $\mathcal{R}$ is in the middle of $\mathcal{O}$ and $\mathcal{C}$ and sees both clocks show
equal times. Therefore the times t_- and t'_- coincide as do τ and τ' and also t_+ and t'_+,
because light from each pair of events in which the clocks show these times reaches
the referee in the same instant.

For the observer $\mathcal{O}$ the event E', in which the moving clock $\mathcal{C}$ shows the time
$\tau' = \tau$, is simultaneous to the event, in which his own clock shows the arithmetic
mean $t = (t_+ + t_-)/2$ of the time t_-, which it shows, when light to E' starts and
the time t_+, at which the reflected light returns. So for $\mathcal{O}$ the event E' occurs at time
t, but the moving clock shows less time, $\tau = \sqrt{t_+ t_-} = \sqrt{1 - v^2}\, t$ (2.15), which is
smaller than the arithmetic mean t (given that the velocity v does not vanish).

For $\mathcal{C}$ the event in which his clock shows the time $t' = (t'_+ + t'_-)/2 = t$ is
simultaneous to the event E, when the clock of $\mathcal{O}$, which moves with respect to $\mathcal{C}$,
shows the time $\tau = \sqrt{t'_+ t'_-} = \sqrt{1 - v^2}\, t$. So for $\mathcal{C}$ the clock of $\mathcal{O}$ runs slower just
as well.

Time dilation is reciprocal because the observers do not agree on which events
are simultaneous. In Euclidean geometry the corresponding fact is commonplace: if
you look in horizontal direction from a lighthouse at sea level to a second lighthouse
of identical construction also at sea level some miles away then the other lighthouse
does not reach the height of the first one because the surface of the earth is curved.
Height depends on which direction is horizontal and the horizontal directions of both
lighthouses do not coincide.

Fig. 2.9 Twins

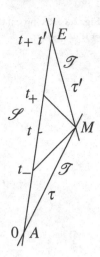

Twin Paradox

Reciprocal time dilation appears to be contradictory, if for example one considers twins. The first twin, the traveller $\mathscr{T}$, departs in an event A with a velocity v for Mars and turns back with velocity v' after the arrival in event M. The other twin, the stay-at-home $\mathscr{S}$, waits calmly the time $t + t'$ until the return of his brother. Which twin, if any, is younger in the end E? For each twin, the other has moved. Does this imply the contradiction, that each twin has aged less than the other?

It is often tacitly insinuated that the observations of both twins agree up to the short acceleration at Mars and that from their observations one cannot distinguish the traveller from the stay-at-home. This is wrong.

Each twin sees the other redshifted during the travel to Mars and blueshifted on the way back. In the first period each twin sees the clock of his brother run slower, in the second faster, than his own clock by a Doppler factor which agrees with the Doppler factor of his brother. But the stay-at-home sees the traveller longer redshifted (and age slower) and shorter blueshifted (and age faster) than the traveller sees the stay-at-home. Both twins see the stay-at-home age more than the traveller.

On arrival at Mars M the traveller $\mathscr{T}$ sees his clock show the travel time τ and a redshifted light ray from the stay-at-home $\mathscr{S}$ show the time t_-, which has elapsed on the clock of $\mathscr{S}$ since the start A. The travel time is larger by a Doppler factor κ (2.1), $\tau = \kappa\, t_-$.

During the return trip the traveller observes on his clock the time τ' go by while blueshifted light shows him that the time $t + t' - t_-$ passes on the clock of the stay-at-home until the end E, $\tau' = \kappa'(t + t' - t_-)$. Altogether, he sees the stay-at-home age by

Fig. 2.10 Equal phases of
acceleration

$$t + t' = \frac{\tau}{\kappa} + \frac{\tau'}{\kappa'} \tag{2.24}$$

while he has grown older by $\tau + \tau'$.

The stay-at-home sees the traveller redshifted during the time $t_+ = \kappa\tau$, which is longer than τ, and blueshifted during the rest of the waiting time $t + t' - t_+ = \kappa'\tau'$, which is shorter than τ'. Altogether, $\mathscr{S}$ ages by

$$t + t' = \kappa\tau + \kappa'\tau' \tag{2.25}$$

while he observes the traveller grow older by $\tau + \tau'$. This agrees with (2.24) as one confirms with (2.13, 2.15) and with the relation $vt + v't' = 0$ that the traveller returns. Both equations imply

$$t + t' = \frac{\tau}{2}\left(\frac{1}{\kappa} + \kappa\right) + \frac{\tau'}{2}\left(\frac{1}{\kappa'} + \kappa'\right) = \frac{\tau}{\sqrt{1 - v^2}} + \frac{\tau'}{\sqrt{1 - v'^2}} . \tag{2.26}$$

The waiting time $t + t'$ is longer than the travel time $\tau + \tau'$.

"Who rests, rusts" and "Travelling keeps young" correctly states the relativistic effect.

The worldline of the traveller differs from the one of the stay-at-home by the acceleration on arrival at Mars M. Such an acceleration is necessary if the second worldline through the two events A and E is to differ from the straight worldline of the first twin because in flat spacetime there is only one straight line through two points.

But the traveller does not become younger during the acceleration. Even if both twins undergo identical phases of acceleration they can age differently, as Fig. 2.10 shows. There both twins travel together until event A where the stay-at-home $\mathscr{S}$ brakes. The traveller $\mathscr{T}$ brakes later to reach M, there he accelerates to return.

After a fitting waiting time the stay-at-home accelerates in exactly the same way and joins the traveller in E from where both continue their joint flight. Between A and E the twins age differently though their acceleration consisted of equal phases. During these phases they age the same, but the remaining pieces of their worldlines constitute the sides of the triangle AME in Fig. 2.9 and there $\mathscr{S}$ ages more.

Time is what clocks show. The clocks of the twins show different times on return. Therefore, time between two events does not only depend on these events but also on the worldline which the clock passes in between; just as in Euclidean geometry the path length between two points of a curve depends on the path which connects both points. Clocks are like mileage counters.

In a spacetime diagram the different aging of the twins is as paradoxical as in Euclidean geometry the statement that in a triangle each side is shorter than the sum of the other two sides. In order to understand triangles one does not need differential geometry of curved spaces, even if one deals with circles and corners, i.e. with curved trajectories. Similarly, the general theory of relativity is not needed for the solution of the twin paradox. It can be used, but gives the same explanation and the same answer as the special theory of relativity: between every two sufficiently adjacent events on the worldline of every free-falling observer there elapses more time than on all other timelike worldlines connecting these two events.

If the two events are not sufficiently adjacent, then gravity can cause the complication that different world lines of free-falling observers connect these events and that on these worldlines different times go by, even though none of the observers has experienced a sensible acceleration. For instance, a space station may orbit the earth in free fall and a second station launched vertically from the earth may fly past the first in free fall during the motion upwards. If the apogee of the second space station is suitably chosen, it can meet the first space station again on the way downwards after the first station has orbited the earth. During the vertical fall more time has elapsed between the two encounters than in the space station orbiting the earth.

The different aging of the twins can be measured with atomic clocks flying around the earth [14] such that for one twin his velocity adds to the revolution of the earth and subtracts for the other twin. In addition, the gravity on ground and during the flight differs and influences the clocks, just as gravity and motion influence the clocks of GPS satellites. There these relativistic effects are routinely accounted for.

Clocks at sea level, which are carried along with the rotating earth, run equally fast. The rotation does not only lead to different velocities, which depend on the longitude, but also to a flattening of the globe, such that the clocks which move faster are further away from the center. Taken together, the different gravity and the different velocity compensate their effects on the clocks at sea level exactly.

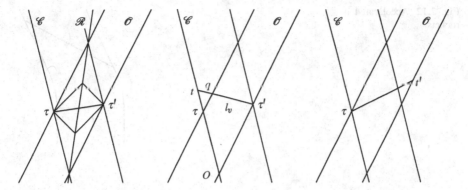

Fig. 2.11 Contraction of moving rods

2.4 Length Contraction

Two moving measuring rods have the same length, if they are equally long for a referee $\mathscr{R}$, who, just as in Fig. 2.11 (left) is always in their middle [23]. The beginning of each rod traverses the worldline of the corresponding observer $\mathscr{C}$ and $\mathscr{O}$, the end traverses a parallel worldline. As the referee $\mathscr{R}$ confirms, the rods of $\mathscr{C}$ and $\mathscr{O}$ have the same length, because in the events τ and τ', which are simultaneous for him and equally far away, both ends of both rods coincide.

A moving rod is shorter than an equal rod at rest by the same factor $\sqrt{1 - v^2}$ by which a moving clock runs slower than an equal clock at rest. This can be deduced from the middle of Fig. 2.11. There we have omitted all auxiliary lines and shown the segment from t to τ' which consists of events which occur simultaneously for the observer $\mathscr{C}$. At this moment, his measuring rod extends from t to τ' and the right ends of both rods coincide. The moving rod is shorter, its left end intersects the line segment from t to τ' in the event q.

The triangles $t\,O\,\tau'$ and $t\,\tau\,q$ are similar, therefore the length l_v of the segment $\tau'q$ relates to the length l of the segment $\tau't$ as the length of $O\tau$ to the length of Ot. But $\tau = \sqrt{1 - v^2}\,t$ is the length of $O\tau$ and t the length of Ot. Therefore, a measuring rod which moves uniformly with a velocity v has the shorter length

$$l_v = \sqrt{1 - v^2}\,l\,,\tag{2.27}$$

if compared to an equal measuring rod of length l at rest.

As the right in Fig. 2.11 shows, length contraction is reciprocal. For the observer $\mathscr{O}$ the events τ and t' occur simultaneous and the measuring rod of $\mathscr{C}$ is shorter.

Fig. 2.12 Accelerated
rockets

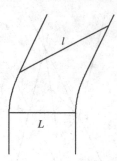

Equally Accelerated Rockets

We consider two rockets which we idealize as points. Initially they rest in a distance L, later they are accelerated in an equal way such that, as in Fig. 2.12, their worldlines are related by a translation by L. For an observer at rest both rockets have a distance L at all times. After the acceleration the rockets follow straight worldlines with velocity v. If the crews of the rockets then measure the mutual distance with measuring rods which they carry along, they obtain some value l. For the observer at rest, this rod is moving and contracted and has length $L = \sqrt{1 - v^2}\, l$, because it reaches from one rocket to the other. So l is larger than L.

A rope as considered in [5, Chap. 9], initially spanned between the rockets and stretched to rupture, snaps immediately, if the rockets and the rope are accelerated equally.

This is also what the crews of both rockets observe. For them the rocket in front reaches the final velocity earlier and veers away from the rear rocket.

If one wants to accelerate the constituents of the rope, which rest initially until a time $t = 0$, to a velocity v, such that their distances, as seen by the constituents, remain unchanged, then one has to accelerate the pieces in the rear more but for a shorter time than the pieces in front such that all points at r, $0 \le r \le L$ traverse worldlines $x(t) = \sqrt{(r + R)^2 + t^2} - R$ during the times $0 \le t \le v(r + R)/\sqrt{1 - v^2}$ and move straight and uniformly afterwards. Here $1/R$ is the acceleration of the last point in the rear.

Length Paradox

Just as time dilation leads to the twin paradox, length contraction seemingly leads to a contradiction, if one considers whether a car with high speed fits into a garage of equal length. For the owner of the garage it is at rest and the moving car is shorter, therefore the car fits into the garage. Seen from the driver, however, the garage is shorter and does not fit the car.

The situation is depicted in the spacetime (Fig. 2.11), where $\mathscr{C}$ and the parallel worldline represent the owner at the gate and the rear wall of the garage while $\mathscr{O}$ and the parallel worldline correspond to the front and rear fender of the car.

Consider a red flash of light, emitted by a photo sensor in the event τ', when the front fender hits the garage wall, and a green flash of light, which is emitted in the event τ, when the rear fender passes the gate. The referee sees both flashes in the same instant and, because he is in the middle of the gate and the wall and the runtimes of light from τ and τ' to him are equal, confirms that the garage and the car have equal length.

The owner of the garage $\mathscr{C}$ observes the red flash from τ' after the green one. If he accounts for the runtime of light, he concludes that the front bumper had hit the garage wall in the event τ' at the time t after the event τ, in which the rear fender passed the gate. For him, the car had fitted into the garage at time t, the car war shorter.

The driver $\mathscr{O}$ sees the green flash of light from the rear of his car τ later than the red flash. If he account for the runtime of light he concludes that the green flash τ had been emitted at the time t' after the red flash. For him, the front fender had hit the wall before the rear fender had passed the gate. So he concludes that the garage is shorter than the car.

This is not a contradiction and not a paradox. Observers, who move relative to each other, do not have to agree on the order of events which are not cause and effect as in the case under consideration. The passage of the rear fender through the gate does not cause the crash of the front fender on the wall and vice versa.

Both observers agree that a fast, slim car can pass a slim garage of equal length if the car in addition has some transverse velocity, just as one can thread long yarn through the narrow eye of a needle.

With some transverse velocity of the car the worldlines of the front and rear fender no longer lie in the plane of Fig. 2.11. They can intersect the plane in the events q and τ. For the garage owner these events are simultaneous, before them the car was on one side and afterwards it is on the other: the car has passed the garage, which is longer than the car. Also the driver observes his car pass the garage, though it is shorter than his car. He first drives around the wall with his front fender and later passes the gate with his rear.

Whether a fast car fits through a garage does not only depend on the length but also on the temporal sequence of the events just as it depends on the direction of a long ladder whether it fits through a low door.

2.5 Doppler Effect

If a clock $\mathscr{C}$ moves with a velocity v in direction $\mathbf{e}$ at an angle θ to the line of sight, then its distance to an observer $\mathscr{O}$ changes by $dr = v \cos\theta dt$ during the short time dt. The changed distance cause a changed runtime of light and light rays $\underline{l}$ and $\bar{l}$ from two events on the clock which started with a time difference dt reach the observer

Fig. 2.13 Doppler effect

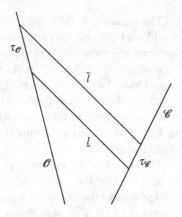

(in units with $c = 1$) with a time difference

$$\tau_{\mathscr{O}} = dt + v \cos\theta \, dt \, . \tag{2.28}$$

In this consideration we use times, velocities and angles as determined by the observer $\mathscr{O}$.

On the moving clock $\mathscr{C}$ the time $\tau_{\mathscr{C}} = \sqrt{1 - v^2} \, dt$ elapses between the emission of the two flashes of light. This follows from (2.13), because spacetime is homogeneous and time flows between the origin $(0, 0, 0, 0)$ and (t, x, y, z) the same as between (t_0, x_0, y_0, z_0) and $(t_0 + dt, x_0 + v_x dt, y_0 + v_y dt, z_0 + v_z dt)$

Consequently the observer $\mathscr{O}$ sees the time

$$\tau_{\mathscr{C}} = \frac{\sqrt{1 - v^2}}{1 + v \cos\theta} \tau_{\mathscr{O}} \tag{2.29}$$

pass by on the moving clock while on his own, equal clock the time $\tau_{\mathscr{O}}$ passes. Equation (2.13), $\tau_{\mathscr{O}} = \kappa \tau_{\mathscr{C}}$, is the special case in which the clock recedes in the line of sight with $\cos\theta = 1$.[2]

If an oscillator is carried along with the clock and oscillates n-times with a frequency $\nu_{\mathscr{C}} = n/\tau_{\mathscr{C}}$, the the observer sees these n oscillations while the time $\tau_{\mathscr{O}}$ passes on his own clock. He observes the frequency $\nu_{\mathscr{O}} = n/\tau_{\mathscr{O}}$,

$$\nu_{\mathscr{O}} = \frac{\sqrt{1 - v^2}}{1 + v \cos\theta} \nu_{\mathscr{C}} \, . \tag{2.30}$$

---

[2] Figure 2.13 depicts the worldlines of the observer $\mathscr{O}$ and the clock $\mathscr{C}$ in a plane. However, we consider the general case in which the worldline of the observer is parallel to the plane and does not intersect the worldline of the clock. Note that in spacetime diagrams the frequency of light is not a property of a light ray but pertains to the distance of two events on parallel light rays.

If $v \cos \theta > \sqrt{1 - v^2} - 1$, then the clock is seen slower and the frequency of light from the clock is shifted to smaller values of red light, it is redshifted.

Otherwise, if $v \cos \theta < \sqrt{1 - v^2} - 1$ and the clock moves towards the observer, then it appears faster and its light is blueshifted. This change of the perceived frequency is the Doppler effect. It is commonly used to measure velocities.

On motion crosswise to the line of sight, $\cos \theta = 0$, the transversal Doppler effect $\tau_{\mathscr{C}} = \sqrt{1 - v^2}\tau_{\mathcal{O}}$ directly shows the slowdown of moving clocks, because the distance between source and observer just does not change.

The Doppler shift is usually time dependent (because the direction changes) and reciprocal only for motion in the line of sight. If the observer $\mathcal{O}$ sends two flashes of light with a delay of $dt = \hat{\tau}_{\mathcal{O}}$ to the clock then the second flash reaches the clock later by $dt' = dt + v \cos \theta dt'$ that is $dt' = \hat{\tau}_{\mathcal{O}}/(1 - v \cos \theta)$. During this interval the time $\hat{\tau}_{\mathscr{C}} = \sqrt{1 - v^2}dt'$ elapses on the moving clock. Seen from the clock, frequencies from $\mathcal{O}$ are shifted to

$$\hat{\nu}_{\mathscr{C}} = \frac{1 - v \cos \theta}{\sqrt{1 - v^2}} \hat{\nu}_{\mathcal{O}} . \tag{2.31}$$

This agrees with $\hat{\nu}_{\mathscr{C}} = \sqrt{1 - v^2}/(1 + v \cos \theta')\hat{\nu}_{\mathcal{O}}$ (2.30) because θ' is the angle to the line of sight, changed by aberration (3.19), with which $\mathscr{C}$ sees $\mathcal{O}$ move.

Apparent Superluminal Velocity

A jet of gas streams out of the quasar 3C273 with a measurable angular velocity [17, Chap. 11]. If one multiplies the observed angular velocity with the known distance one obtains a velocity of seven times the speed of light for the crosswise motion. The quasar seems to emit particles with superluminal velocity.

This conclusion is wrong, the product of the distance with the observed angular velocity is not the velocity transverse to the line of sight.

The clock $\mathscr{C}$ in Fig. 2.13 moves by $v \sin \theta \, dt = r \, d\theta$ within the short time dt transverse to the line of sight, where r denotes its present distance. The flashes of light $\underline{l}$ and $\bar{l}$ reach the observer with a difference angle $d\theta$ and a time difference $\tau_{\mathcal{O}} = dt + v \cos \theta \, dt$ because $\bar{l}$ starts later by dt and has to pass a distance which is larger by $dr = v \cos \theta \, dt$. So the observed angular velocity $\omega_{\mathcal{O}} = d\theta/\tau_{\mathcal{O}}$ and the apparent transverse velocity $u = r \, \omega_{\mathcal{O}}$ are

$$\omega_{\mathcal{O}} = \frac{v \sin \theta}{r(1 + v \cos \theta)} , \quad u = \frac{v \sin \theta}{1 + v \cos \theta} . \tag{2.32}$$

This velocity u becomes maximal for the angle $\cos \theta = -v$ between the direction of motion and the line of sight and in this case has the value $v/\sqrt{1 - v^2}$. This value can be arbitrary large though $|v|$ is smaller than $c = 1$, the speed of light.

Fig. 2.14 Spherical coordinates

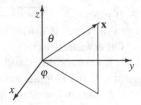

2.6 Spacetime Coordinates

We can denote the events E in spacetime simply by the values[3] $(t_+, t_-, \theta, \varphi)$, the light coordinates of E, which an observer $\mathcal{O}$ determines as he sends light to E and receives it from E. He reads the times of emission, t_-, and reception, t_+, from his clock and determines the direction of the outgoing light ray by means of, say, two angles θ and φ.

Primarily coordinates only have to denote the events uniquely, at least in some range of their values. Other coordinates, which are invertible functions of the light coordinates, are equally conceivable. In particular, light coordinates are related in a simple way to inertial coordinates (t, x, y, z), in which particles, which move uniformly on straight lines, traverse straight coordinate lines (Fig. 2.14).

The time t and the distance $r = \sqrt{x^2 + y^2 + z^2}$, at which the event E occurs, are the arithmetic mean and half the difference of the light coordinates t_+ and t_- (1.4, 1.5),

$$t = \frac{t_+ + t_-}{2} , \quad r = \frac{t_+ - t_-}{2} . \tag{2.33}$$

The direction of the outgoing light ray from $\mathcal{O}$ to the event E is opposite to the incident direction of the light ray from E because the observer does not rotate but uses reference directions that do not change in time.

The angles θ and φ of the light ray to E and the distance r are the spherical coordinates and define the cartesian spatial coordinates of the event by

$$\mathbf{x} = \begin{pmatrix} x \\ y \\ z \end{pmatrix} = r\mathbf{e}_{\theta,\varphi} = r \begin{pmatrix} \sin\theta\cos\varphi \\ \sin\theta\sin\varphi \\ \cos\theta \end{pmatrix} . \tag{2.34}$$

For events on the worldline of the observer $\mathcal{O}$ one has $t_+ = t_-$, hence $\mathbf{x} = 0$. In particular, the origin O has coordinates $(0, 0, 0, 0)$.

If $\mathcal{O}$ emits a light ray at time t_0 in the direction $\mathbf{e}_{\theta,\varphi}$, then the light ray passes events for which t_-, θ and φ are constant

$$t = \frac{t_+ + t_0}{2} , \quad \mathbf{x}(t) = \frac{t_+ - t_0}{2}\mathbf{e}_{\theta,\varphi} , \tag{2.35}$$

---

[3] θ and φ are the Greek letters theta and phi.

or, if we express the variable t_+ in terms of t, then the light ray is given by the map

$$\Gamma : t \mapsto (t, \mathbf{x}(t)) = (t, \mathbf{e}_{\theta,\varphi} \cdot (t - t_0)) \,. \tag{2.36}$$

This is a worldline parameterized by t, which at the time t_0 intersects the worldline of the observer. In the coordinates (t, x, y, z) it is a straight worldline which is traversed with the speed of light $c = 1$, since the speed $\mathbf{v} = \frac{d\mathbf{x}}{dt}$ is a unit vector. The equations also holds for $t < t_0$ for a light pulse incident from the opposite direction $-\mathbf{e}_{\theta,\varphi}$. For such a light ray $t_+ = t_0$ is constant and $t = (t_0 + t_-)/2$ and $\mathbf{x} = -\mathbf{e}_{\theta,\varphi}(t_0 - t)$. While light coordinates $(t_+, t_-, \theta, \varphi)$ of a passing light ray are discontinuous in the event, in which it intersects the worldline of the observer, inertial coordinates are continuous.

Displacing the light ray by $\mathbf{x}_0 + \mathbf{e}_{\theta,\varphi}t_0$ yields more generally the light ray which passes $\mathbf{x}_0$ at the time $t = 0$,

$$\Gamma : t \mapsto (t, \mathbf{x}(t)) = (t, \mathbf{e}_{\theta,\varphi} \cdot t + \mathbf{x}_0) \,. \tag{2.37}$$

If the worldline of a linearly and uniformly moving particle passes the origin O at time $t = 0$, the observer $\mathscr{O}$ sees afterwards all events on this worldline from the same direction. The angles θ and φ are constant, except at $t = 0$. The particle departs into the opposite of the direction from which it approached and the angles change discontinuously from θ to $= \pi - \theta$ and from φ to $\varphi + \pi$ at $t = 0$.

According to (2.11) one has $t_+ = \kappa^2 t_-$ for events on the straight worldline of the particle. For its coordinates this means

$$t = (\kappa^2 + 1)\frac{t_-}{2} \,, \quad \mathbf{x} = (\kappa^2 - 1)\frac{t_-}{2}\mathbf{e}_{\theta,\varphi} \,, \tag{2.38}$$

or, if we express t_- in terms of t and use (2.14), the worldline is given by

$$\Gamma : t \mapsto (t, \mathbf{x}(t)) = (t, \mathbf{v}\,t) \quad \text{with} \quad \mathbf{v} = \frac{d\mathbf{x}}{dt} = \frac{\kappa^2 - 1}{\kappa^2 + 1}\mathbf{e}_{\theta,\varphi} \,. \tag{2.39}$$

Translating the worldline by $\mathbf{x}_0$ one obtains more generally the worldline of an uniformly moving particle which passes the point $\mathbf{x}_0$ at time $t = 0$,

$$\Gamma : t \mapsto (t, \mathbf{x}(t)) = (t, \mathbf{v}\cdot t + \mathbf{x}_0) \,. \tag{2.40}$$

So the coordinates (t, x, y, z) which we have constructed from the light coordinates t_+, t_-, θ and φ are inertial coordinates in which particles, which move straight and uniformly, traverse straight coordinate lines.

2.7 Scalar Product and Length Squared

Together, the time and the spatial coordinates of each event constitute an ordered set (t, x, y, z) of four real numbers, and each such four-tuple corresponds to one and only one event. In Special Relativity spacetime, the set of all events, is $\mathbb{R}^4$. We denote the components of the four-tuple which corresponds to a particular event E either by t_E, x_E, y_E and z_E or we use a name like u for the four-tuple $u = (u^0, u^1, u^2, u^3)$ and enumerate the components with a superscript. It depends on the context whether the superscript denotes an exponent (rarely) or enumerates a footnote or a component.

The homogeneity of spacetime makes it a vector space. If one shifts all events $u = (u^0, u^1, u^2, u^3)$, which participate in some physical process, by $s = (s^0, s^1, s^2, s^3)$ in space and time, then the events

$$u + s = (u^0 + s^0, u^1 + s^1, u^2 + s^2, u^3 + s^3) \qquad (2.41)$$

can participate in an equally possible process.

The scaled versions of spacetime diagrams consist of events

$$au = (au^0, au^1, au^2, au^3) \qquad (2.42)$$

scaled by a common factor a. However, elementary physical processes are not scale invariant. While scaled diagrams of physical processes with free pointlike particles correspond to equally possible physical processes, this is not true for interacting particles, for example, one has never observed an enlarged hydrogen atom (electron and proton bound by electromagnetic interactions).

The set $\mathbb{R}^4$, equipped with the operations of addition and multiplication by a scale factor, is a four-dimensional vector space. Its elements are called four-vectors.[4]

On an uniformly moving clock, which passes the two events (t_0, x_0, y_0, z_0) and $(t_0 + t, x_0 + x, y_0 + y, z_0 + z)$, there elapses the time

$$\tau^2 = t^2 - x^2 - y^2 - z^2 . \qquad (2.43)$$

This follows from (2.13), because spacetime is homogeneous and time flows between the origin $(0, 0, 0, 0)$ and the event (t, x, y, z) the same as between (t_0, x_0, y_0, z_0) and $(t_0 + t, x_0 + x, y_0 + y, z_0 + z)$.

The time between two events does not depend on the details of the clock used to measure it. The time is a measure for distance, i.e. a geometric structure, in spacetime.

---

[4] Without mentioning it explicitly we shall consider different copies of $\mathbb{R}^4$, e.g. spacetime or the set of four-velocities, four-momenta or four-accelerations. Vectors from different spaces cannot be added, because they differ in units. e.g. a velocity $\mathbf{v}$ cannot be added to a position $\mathbf{x}$. What can be added is the image $\mathbf{v}t$ of a velocity $\mathbf{v}$ under the linear map t, which maps it to the space of positions. Though vectors from different four-spaces cannot be added, their directions can be compared, because, as we shall see, the Lorentz group acts on each of these spaces and the x-direction, for example, is the set of vectors which is invariant under rotations around the x-axis and under boosts in y- and z-directions (also compare page 89).

Events with equal temporal distance one does not find in a plane $t = $ const as in nonrelativistic physics or on a sphere $x^2 + y^2 + z^2 = r^2 = $ const as in Euclidean geometry, but on an hyperboloid $t^2 - x^2 - y^2 - z^2 = \tau^2 = $ const. The square of the temporal distance between two events is not subject to the Pythagorean theorem but to the theorem of Minkowski.

The clock does not depend on which observer determines coordinates for the events. If another observer measures light coordinates t'_+, t'_-, θ' and φ' and converts them into spacetime coordinates (t'_0, x'_0, y'_0, z'_0) and $(t'_0 + t', x'_0 + x', y'_0 + y', z'_0 + z')$ of the two events, then the sums of squares appearing in (2.43) have to agree

$$t^2 - x^2 - y^2 - z^2 = t'^2 - x'^2 - y'^2 - z'^2 \,. \tag{2.44}$$

The sum of squares plays a central role in relativistic physics. We introduce the related scalar product of four-vectors like $u = (u^0, u^1, u^2, u^3)$ and $w = (w^0, w^1, w^2, w^3)$

$$u \cdot w := u^0 w^0 - u^1 w^1 - u^2 w^2 - u^3 w^3 \,. \tag{2.45}$$

As length squared of a four-vector w we define[5]

$$w^2 = w \cdot w = (w^0)^2 - (w^1)^2 - (w^2)^2 - (w^3)^2 \,. \tag{2.46}$$

In this notation, the time τ between events u and w is given by

$$\tau^2 = (u - w)^2 \,. \tag{2.47}$$

The scalar product (2.45) maps each pair of four-vectors to a real number and is symmetric and linear in each argument (a denotes an arbitrary real factor)

$$v \cdot w = w \cdot v \,, \tag{2.48}$$

$$u \cdot (v + w) = u \cdot v + u \cdot w \,, \quad v \cdot (a w) = a (v \cdot w) \,, \tag{2.49}$$

but, different from Euclidean geometry, not definite. Lightlike vectors have length squared zero though they do not vanish. The scalar product is nondegenerate, i.e. the scalar product of a vector v vanishes with all other vector if and only if $v = 0$ vanishes.

The scalar product of two vectors u and v can be written as the difference of lengths squared

$$u \cdot v = \frac{1}{4}((u + v)^2 - (u - v)^2) \,. \tag{2.50}$$

Since different observers determine different coordinates but the same lengths squared of differences of four-vectors (2.44), scalar products of difference vectors

---

[5] The reader has to deduce from the context whether the length squared or the y-component of a vector is meant.

do not depend on the coordinate system of the respective observer either,

$$u \cdot v = u' \cdot v' . \tag{2.51}$$

If the length squared w^2 is positive we call w timelike, if it is negative we call w spacelike, if $w^2 = 0$, $w \neq 0$, w is called lightlike. A timelike or lightlike vector w is future directed, if its component w^0 is positive, otherwise it is past directed.

Two events A and B are mutually spacelike if the corresponding difference vector from B to A

$$w_{AB} = (t_A - t_B, x_A - x_B, y_A - y_B, z_A - z_B) \tag{2.52}$$

is spacelike. Correspondingly we define lightlike or timelike pairs of events.

An event B can cause an effect A only, if w_{AB} is future directed timelike or lightlike.

Events on a light ray are mutually lightlike.

Events on the worldline of an observer are mutually timelike since each observer is slower than light. If his worldline is straight then the length squared of the difference vector of two of his events is the square of the time which passes on his clock between the two events.

Orthogonal

To construct the line $\mathcal{O}_\perp$, which orthogonally intersects the line $\mathcal{O}$ in the point t, one chooses two points on $\mathcal{O}$, t_+ and t_-, which are equally far away from t, and determines a second point E which is equally far away from t_+ and t_-. The orthogonal line $\mathcal{O}_\perp$ is the line through t and E.

This is true in Euclidean geometry and in spacetime. In spacetime, however, the distance is given by τ (2.43). If t_- and t_+ are two events on the worldline of the observer $\mathcal{O}$ and if t is in their middle then the intersections E and E' of the light rays through t_- and t_+ lie on the orthogonal line through t, because E and E' are equally far away from t_- and from t_+, to wit the distance vanishes because the separations are lightlike (Fig. 2.15).

The worldline $\mathcal{O}$ consists of events which are equilocal for the observer. The events on $\mathcal{O}_\perp$ are equitemporal for him (Fig. 1.6). The lines of equilocal events are orthogonal to the lines of equitemporal events.

Using the vector v from t_- to t and from t to t_+ and the vector w from t to E, the light ray from t_- to E is $v + w$. The vector $v - w$ is the light ray back from E to t_+. The length squared of the lightlike vectors $v + w$ and $v - w$ vanishes,

$$\begin{aligned} 0 &= (v + w)^2 = v^2 + 2v \cdot w + w^2 , \\ 0 &= (v - w)^2 = v^2 - 2v \cdot w + w^2 . \end{aligned} \tag{2.53}$$

Fig. 2.15 Orthogonal vectors
with hyperbola

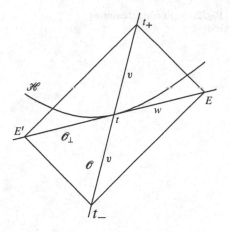

Therefore $v^2 = -w^2$ and the scalar product of the orthogonal vectors vanishes,

$$v \cdot w = 0 \, . \tag{2.54}$$

The length squared v^2 is the square of the time between the events t_- and t on the worldline of the observer $\mathcal{O}$. This is the runtime of light and therefore the distance from $\mathcal{O}$ to the event E. Because $v^2 = -w^2$ the square of the distance of two simultaneous events, which are separated by the spacelike vector w, is $-w^2$.

Orthogonal: *The vector w from an event t on the worldline of a uniformly moving observer to an event E, which occurs simultaneously for him, is orthogonal in terms of the scalar product (2.45) to his worldline. The negative length squared $-w^2$ is the square of the distance between E and the observer.*

The hyperbola $\mathcal{H}$ through t around t_- is defined to consist of points which are obtained from t_- by equally long translations $u(s)$

$$u(s) = \sqrt{1 + s^2}\, v + s w \, , \quad u(s)^2 = v^2 \, , \tag{2.55}$$

where s varies in the real numbers. In particular, the point t on the hyperbola corresponds to $s = 0$. In terms of the length squared of spacetime, all points of $\mathcal{H}$ are equally far away from t_-.

Each vector from t_- to a point A on the orthogonal line $\mathcal{O}_\perp$ is of the form $x(s) = v + sw$, where s is some real number. Because of $v \cdot w = 0$ it is as long as the vector $-v + sw$ from t_+ to A.

Because $\sqrt{1 + s^2} > 1$ for $s \neq 0$, all points of $\mathcal{H}$ apart from t lie on the side of $\mathcal{O}_\perp$ which is opposite to t_-, one has $u(s) = x(s) + a(s)v$ with a positive $a(s)$. In addition, t belongs to both $\mathcal{O}_\perp$ and $\mathcal{H}$. Therefore both curve touch each other at t and the straight line $\mathcal{O}_\perp$ is tangent to the hyperbola $\mathcal{H}$ in the point t. The tangent at t is orthogonal to the vector from t_- to t.

Fig. 2.16 Rotated measuring rods

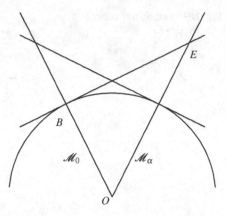

The same conclusion is obtained by differentiating $u(s)^2$ with respect to s. The tangential vector $t(s) = \frac{du}{ds}(s)$ is orthogonal to the position vector $u(s)$,

$$u(s) \cdot u(s) = \text{constant} \Rightarrow \frac{du}{ds} \cdot u = 0 . \tag{2.56}$$

2.8 Perspectives

If one takes bearing in horizontal direction from a lighthouse at sea level to a second lighthouse of identical construction also at sea level some miles away then the other lighthouse does not reach the same height because the surface of the earth is curved (see Fig. 2.16). Height is a perspective quantity. It depends on which direction is horizontal and the horizontal directions of both lighthouses do not coincide.

Perspective shortening is physically relevant, because one can change the height of a ladder by rotation, it may pass a low door though the ladder is longer than the height of the door and though rotations leave the sizes of the door and the ladder unchanged.

Figure 2.16 depicts the perspective height of two measuring rods $\mathcal{M}_0$ and $\mathcal{M}_\alpha$ in Euclidean geometry which are rotated with respect to each other. The circle consists of points of equal distance to the center; each tangent vector is orthogonal to the position vector.

The measuring rods intersect in the point O. For an observer, who measures height with $\mathcal{M}_0$, all point on the straight line through B, which is orthogonal to $\mathcal{M}_0$, are equally high. In particular the point E is as high as B and higher as the end point of $\mathcal{M}_\alpha$. Rotated measuring rods reach less high.

Fig. 2.17 Time dilation

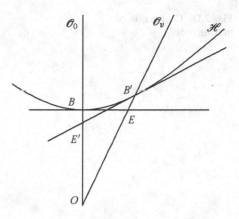

The perspective shortening of height is reciprocal. Judged from $\mathcal{M}_\alpha$ the rotated rod $\mathcal{M}_0$ is lower.

If one replaces in Fig. 2.16 the circle by a hyperbola, one obtains the geometric relations in spacetime.

In Fig. 2.17 the hyperbola $\mathcal{H}$ consists of events with equal temporal distance τ to the origin O (2.43). Equal uniformly moving clocks of observers $\mathcal{O}_0$ and $\mathcal{O}_v$, who pass the origin, show the same laps of time, τ, when their worldlines intersect the hyperbola.

The tangents in B and B' are orthogonal to the worldlines of the observers $\mathcal{O}_0$ and $\mathcal{O}_v$ respectively (2.56). Therefore they consist of events which occur simultaneously for $\mathcal{O}_0$ or $\mathcal{O}_v$. The tangents intersect the worldline of the other observer before the time τ has elapsed on it.

If the time τ elapses on a clock between two events, then the shorter time $\tau_{EO} = \tau_{E'O} = \sqrt{1 - v^2}\tau$ (2.13) passes between the simultaneous events on a moving clock, just as two points on a vertical ladder have a shorter distance than equally high points on a tilted ladder. The perspective relations in spacetime are reciprocal as in Euclidean geometry.

The abbreviated summary "moving clocks run slower" suppresses the specifications of the segments OE and OB or OE' and OB' the duration of which is to be compared. The abbreviation is the reason for misunderstandings, because "running slower" is an order relation and the clock of $\mathcal{O}_0$ cannot run slower and also faster than the clock of $\mathcal{O}_v$. In fact, both clocks are equal as confirmed by the referee in Fig. 2.2.

Contraction of moving measuring rods can be read of Fig. 2.18 which is the mirrored version of Fig. 2.17. The beginning and the end of uniformly moving measuring rods of observers $\mathcal{O}_0$ and $\mathcal{O}_v$ traverse pairs of parallel straight worldlines. Both rods have the length l, because the left ends coincides in O and the right ends B and B' lie on the auxiliary hyperbola, which consists of points P which satisfy $-w_{PO}{}^2 = l^2$.

Fig. 2.18 Contraction of
moving rods

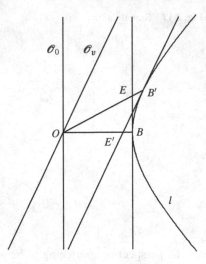

Fig. 2.19 Twin paradox

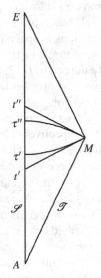

The segments OB and OB' are orthogonal to the worldlines of the observers and $\mathcal{O}_v$, because they are position vectors and parallel to tangent vectors of the hyperbola (page 43).

Therefore the events O, E' and B are simultaneous for $\mathcal{O}_0$. At this moment, the left ends coincide, but the right end of the moving rod only reaches to E', so the moving rod is shorter than the own rod which reaches until B.

For $\mathcal{O}_v$ the events O, E and B' are simultaneous. Then the left ends coincide in O and the rod, which moves relative to $\mathcal{O}_v$ only reaches to E and is shorter than the own rod which reaches until B'.

In Fig. 2.19 the waiting time and the travel times of the twin paradox are not determined from the Doppler factors as in Fig. 2.9 but compared by the auxiliary hyperbola from M to τ' with origin at A and the hyperbola from M to τ'' with origin at E.

The line segments from A to τ' and from τ'' to E on the worldline of the stay-at-home $\mathscr{S}$ last as long as for the traveller $\mathscr{T}$ the travel to and from Mars. The stay-at-home has aged in addition during the time which passed between τ' and τ''. On the straight worldline of the stay-at-home more time has passed than on the worldline of the traveller with a kink.

The tangents Mt' and Mt'' of the hyperbolas consist of events, which are simultaneous to the arrival at Mars for an uniformly moving observer $\mathscr{O}_{\text{to}}$, who also flies to Mars respectively for an uniformly moving observer $\mathscr{O}_{\text{from}}$, who flies back. The tangents intersect the worldline of $\mathscr{S}$ in t' and t'' confirming that for observers who fly to or from Mars the clock of the stay-at-home shows less time than simultaneously on their own clocks.

But the events t' and t'' do not coincide, they are simultaneous to the arrival at Mars for different observers. Between t' and t'' the stay-at-home ages so much that in the end he has grown older than the traveller.

Chapter 3
Transformations

Abstract Motion of an observer makes him measure Doppler shifted or Lorentz transformed light coordinates. These Lorentz transformations relate the velocities of particles which different observers measure, and change by aberration the directions, from which incident light rays are perceived. Aberration is conformal: angles and relative sizes of small, neighbouring objects agree in aberrated pictures. Lorentz transformations determine how the energy and the momentum of a particle depend on its velocity. The conservation of energy and momentum restricts the decay of a particle and the scattering of two particles with observable consequences.

3.1 Lorentz Transformation of Coordinates

How do the coordinates (t, x, y, z) which an observer $\mathcal{O}$ attributes to an event E correspond to the coordinates (t', x', y', z') which an observer $\mathcal{O}'$ measures for the same event, if he moves with a velocity v relative to $\mathcal{O}$?

To begin with we investigate the case that the worldlines of the two observers intersect in the event O and that they set their clocks to $t' = t = 0$ on this occasion. To keep the discussion simple, both observers choose their x-axis in the direction of relative motion such that for $\mathcal{O}$ the observer $\mathcal{O}'$ moves in x-direction and vice versa $\mathcal{O}$ moves in the $-x'$-direction of $\mathcal{O}'$. Then the y- and z-coordinates of each event E in the plane of the worldlines of the observers vanish, $y = z = 0 = y' = z'$.

Each observer sees the clock of the other observer slowed down by the same Doppler factor $\kappa(\mathcal{O}, \mathcal{O}') = \kappa(\mathcal{O}', \mathcal{O})$ (2.10). When flashes of light to and from an event E as in Fig. 3.1 pass the observers, then the times on their clocks t'_- and t_- and also t_+ and t'_+, are proportional to each other with the Doppler factor (2.13)

$$\kappa(v) = \sqrt{\frac{1+v}{1-v}} = \frac{1+v}{\sqrt{1-v^2}} \tag{3.1}$$

N. Dragon, *The Geometry of Special Relativity—a Concise Course*,
SpringerBriefs in Physics, DOI: 10.1007/978-3-642-28329-1_3,
© The Author(s) 2012

Fig. 3.1 Lorentz transformation

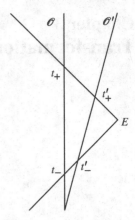

because they are pairs of times which equal clocks show at the emission and reception of a flash of light. So the observer $\mathscr{O}'$ measures the light coordinates

$$t'_+ = \kappa^{-1} t_+ , \quad t'_- = \kappa t_- . \tag{3.2}$$

The transformation of the two light coordinates decomposes into two transformations of coordinates, which scale separately—t'_+ depends only on t_+ and t'_- only on t_-—and inversely, such that their product, the area $t_+ t_- = t'_+ t'_-$ of the lightangle with corners O and E, remains unchanged.

In spacetime coordinates $t' = (t'_+ + t'_-)/2$ and $x' = (t'_+ - t'_-)/2$ (1.4, 1.5) the transformation is coupled

$$t' = \frac{1}{2}(\kappa + \kappa^{-1})t - \frac{1}{2}(\kappa - \kappa^{-1})x , \quad x' = -\frac{1}{2}(\kappa - \kappa^{-1})t + \frac{1}{2}(\kappa + \kappa^{-1})x . \tag{3.3}$$

Inserting $\kappa(v)$ and $1/\kappa(v) = \kappa(-v) = (1-v)/\sqrt{1-v^2}$ (2.16) one obtains

$$t' = \frac{t - vx}{\sqrt{1-v^2}}, \quad x' = \frac{-vt + x}{\sqrt{1-v^2}} \quad \text{or} \quad \begin{pmatrix} t' \\ x' \end{pmatrix} = \frac{1}{\sqrt{1-v^2}} \begin{pmatrix} 1 & -v \\ -v & 1 \end{pmatrix} \begin{pmatrix} t \\ x \end{pmatrix}. \tag{3.4}$$

This is (in units with $c = 1$) the Lorentz transformation of the coordinates of an event in the t–x-plane to the t'–x'-coordinaten which an observer, who moves in x-direction with velocity v, attributes to the same event. Such a coordinate transformation is called passive. It does not change the events. A transformation is called active, if it transforms events. If in (3.4) one changes the sign of v, then one obtains the active Lorentz transformation which maps the worldline of a particle at rest to the worldline of a particle, which moves with velocity v in x-direction.

From (3.2) one concludes that κ^{-1} and therefore the negative velocity $-v$ corresponds to the inverse transformation.

Lorentz Transformation in Four Dimensions

Also if the event E does not lie in the t–x-plane, the Lorentz transformation Λ of the (t, x, y, z)-coordinates to (t', x', y', z')-coordinates has to be linear because the worldlines of free particles are straight lines for each observer. Consequently, triangles formed by three intersecting straight lines, are mapped to triangles. The sides of a triangle correspond to vectors u and v and, for the third side, $w = u+v$. They are mapped to $\Lambda(u)$, $\Lambda(v)$ and $\Lambda(u+v) = \Lambda(u)+\Lambda(v)$, because triangles transform into triangles. From $\Lambda(u + u) = \Lambda(u) + \Lambda(u)$ one concludes $\Lambda(nu) = n\Lambda(u)$ for integer n, and from $\Lambda(u) = \Lambda(n(1/nu)) = n\Lambda(1/nu)$ one deduces $\Lambda(au) = a\Lambda(u)$ for rational a and because Λ is continuous for each real a. Therefore Λ is linear.

The coordinates y' and z' are linear combinations of t, x, y and z, which vanish for arbitrary t and x if the event E lies in the plane $y = z = 0$. So y' and z' do not depend on t and x

$$\begin{pmatrix} y' \\ z' \end{pmatrix} = \begin{pmatrix} a & b \\ c & d \end{pmatrix} \begin{pmatrix} y \\ z \end{pmatrix}. \tag{3.5}$$

Also $t' = (t - vx)/\sqrt{1 - v^2} + ey + fz$ and $x' = (-vt + x)/\sqrt{1 - v^2} + gy + hz$ are the most general linear combinations which coincide with (3.4) for $y = z = 0$.

The Lorentz transformation has to leave scalar products invariant (2.51). Therefore e, f, g and h vanish: the vectors $(\sqrt{1 - v^2}, 0, 0, 0)$, $(0, \sqrt{1 - v^2}, 0, 0)$, $(0, 0, 1, 0)$ and $(0, 0, 0, 1)$ are mutually orthogonal. For the observer $\mathcal{O}'$ they have components $(1, -v, 0, 0)$, $(-v, 1, 0, 0)$, (e, g, a, c) and (f, h, b, d) and the first two vectors have to be orthogonal to the last two.

So (3.4) and (3.5) are valid for arbitrary (t, x, y, z), where (a, c) and (b, d) are the components of Lorentz transformed normalized, mutually orthogonal vectors and consequently are normalized and mutually orthogonal. By $a^2 + c^2 = 1$ the coefficients a and c can be written as cosine and sine of some angle α. Because of $ab + cd = 0$ the coefficients (b, d) are a multiple of $(-c, a)$, and due to $b^2 + d^2 = 1$ only $(b, d) = \pm(-c, a)$ can hold. The Lorentz transformation of the plane orthogonal to the direction of motion is a rotation (or a rotary reflection)

$$\begin{pmatrix} y' \\ z' \end{pmatrix} = \begin{pmatrix} \cos \alpha & -\sin \alpha \\ \sin \alpha & \cos \alpha \end{pmatrix} \begin{pmatrix} y \\ z \end{pmatrix}. \tag{3.6}$$

If the angle of rotation vanishes, $\alpha = 0$, the Lorentz transformation is called a boost or rotation-free,

$$t' = \frac{t - vx}{\sqrt{1 - v^2}}, \quad x' = \frac{-vt + x}{\sqrt{1 - v^2}}, \quad y' = y, \quad z' = z. \tag{3.7}$$

In matrix notation the transformation reads

Fig. 3.2 Lorentz flow

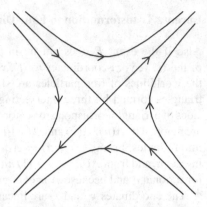

$$\begin{pmatrix} t' \\ x' \\ y' \\ z' \end{pmatrix} = \begin{pmatrix} \frac{1}{\sqrt{1-v^2}} \begin{pmatrix} 1 & -v \\ -v & 1 \end{pmatrix} & \\ & 1 \\ & & 1 \end{pmatrix} \begin{pmatrix} t \\ x \\ y \\ z \end{pmatrix} . \tag{3.8}$$

If one continuously increases the angle α of a rotation D_α around some axis from zero to some value $\overline{\alpha}$ then as a function of α a point $D_\alpha x = x(\alpha)$, the orbit of $x(0)$ under rotations, traverses a segment of a circle, because rotations leave Euclidean distances invariant. By the same reason points in the Fig. 3.2 orbit parts of hyperbolas, if one applies Lorentz transformations Λ_v and varies the velocities v. Lightangles with the origin as one corner and a point on the hyperbola as opposite corner have the same area, $t_+ t_- = t'_+ t'_-$. For Lorentz transformations leave the length squared $t^2 - x^2 - y^2 - z^2$ invariant. Lightlike vectors get stretched or shrunken. The origin is a hyperbolic fixed point. It is a stagnation point of the flow, not a vortex as in the case of rotations.

If we denote in (3.8) the four-vector (t, x, y, z) just by x and the 4×4 Lorentz matrix by Λ, then the transformation takes the form

$$x' = \Lambda x . \tag{3.9}$$

Vice versa one has $x = \Lambda^{-1} x'$. The coordinates (t, x, y, z) are obtained from (t', x', y', z') by multiplication with the inverse Lorentz matrix. For a rotation-free Lorentz transformation this is the original matrix in which the velocity v is replaced by $-v$, $\Lambda_v^{-1} = \Lambda_{-v}$.

More generally the observer $\mathcal{O}'$ can move with velocity $\mathbf{v}$ in an arbitrary direction and his reference directions can be rotated with respect to $\mathcal{O}$. The matrix of such a Lorentz transformation is of the form $\Lambda = D_1 \Lambda_v D_2$ (6.35), where D_1 and D_2 denote rotation matrices and Λ_v is the matrix in (3.8) which corresponds to motion in x-direction.

More generally the worldlines of the two observers do not have to intersect but can be shifted in space and time by $a = (a^0, a^1, a^2, a^3)$. Then the x'-coordinates are related to the x-coordinates by a Poincaré transformation $T_{\Lambda,a}$

$$T_{\Lambda,a} : \begin{cases} \mathbb{R}^4 \to \mathbb{R}^4 \\ x \mapsto x' = \Lambda x + a \end{cases} \tag{3.10}$$

Two Poincaré transformations T_{Λ_1,a_1} and T_{Λ_2,a_2}, carried out successively, yield a further Poincaré transformation, their product

$$T_{\Lambda_2\Lambda_1,\Lambda_2 a_1+a_2} = T_{\Lambda_2,a_2} \circ T_{\Lambda_1,a_1} . \tag{3.11}$$

The inverse transformation is $\left(T_{\Lambda,a}\right)^{-1} = T_{\Lambda^{-1},-\Lambda^{-1}a}$. Poincaré transformations constitute a group (page 77), i.e. a set of elements g with an associative product and a unit element where each element has an inverse g^{-1}.

Poincaré transformations are not the most general transformations, which map lightcones to lightcones and leave invariant the velocity of light. It is invariant under the larger group of conformal transformations which contains in particular dilations $x \mapsto e^a x$ by an invertible factor e^a.

Addition of Velocities

Let the worldline $\Gamma : \lambda \mapsto x(\lambda)$ of a particle be parameterized such that the time t strictly increases with the parameter λ, $dt/d\lambda > 0$. The tangent vector, the derivative $dx/d\lambda$, transforms under Poincaré-transformations simply as a four vector and for an observer, moving with a velocity u, $|u| < 1$, has components, which are related by

$$\frac{dx'}{d\lambda} = \Lambda \frac{dx}{d\lambda} \tag{3.12}$$

to the components for an observer at rest. The velocity $\mathbf{v}$, the derivative of the spatial components with respect to the time t

$$\frac{d\mathbf{x}}{d\lambda} = \frac{d\mathbf{x}}{dt}\frac{dt}{d\lambda} = \mathbf{v}\frac{dt}{d\lambda} \tag{3.13}$$

is a rational function of the derivatives with respect to the parameter λ. Therefore, the velocity $\mathbf{v}'$, which the moving observer measures, is a rational function of the components of $\mathbf{v}$, $(i, j, k \in \{1, 2, 3\})$

$$v'^i = \frac{\Lambda^i{}_m \dfrac{dx^m}{d\lambda}}{\Lambda^0{}_n \dfrac{dx^n}{d\lambda}} = \frac{\Lambda^i{}_j v^j + \Lambda^i{}_0}{\Lambda^0{}_0 + \Lambda^0{}_k v^k} . \tag{3.14}$$

In particular, if we choose the x-axis in direction of the relative motion (3.8) and the y-axis such that $\mathbf{v}$ lies in the x–y-plane and if we denote $\mathbf{v}$ by its modulus and its angle θ

$$(v_x, v_y, v_z) = v(\cos\theta, \sin\theta, 0) , \quad \text{and} \quad (v_x', v_y', v_z') = v'(\cos\theta', \sin\theta', 0) . \quad (3.15)$$

then the transformation law of **v**, solved for $\cos\theta'$ and v' yields

$$\cos\theta' = \frac{v\cos\theta - u}{\sqrt{(u - v\cos\theta)^2 + (1 - u^2)v^2\sin^2\theta}} , \quad (3.16)$$

$$v' = \frac{\sqrt{(u - v\cos\theta)^2 + (1 - u^2)v^2\sin^2\theta}}{1 - u\,v\cos\theta} . \quad (3.17)$$

3.2 Perception

If two mutually moving observers meet in some event, which we choose as origin O, then their past lightcones coincide and each light ray seen by one observer is also perceived by the other observer. Nevertheless they see something different. Color, direction and luminosity of the light depend on the velocity of the observer relative to the source.

The Doppler effect changes the frequency, i.e. the color, of the light and in addition the number of photons received per second. Aberration changes the direction **n**, from which the light ray is seen to come. The change of their directions changes also the density of the light rays.

Hence, the moving observer sees a smoothly deformed version of the image of the stationary observer with changed colors and brightness. Topologically, i.e. in terms of the relative positions the two images coincide, since they are related by a continuous invertible map. In particular, a moving observer cannot see things hidden from an observer at rest in this instant at this position. However, at high speed the moving observer's field of vision in the direction of motion contains light rays which for the stationary observer come from behind [12].

To be definite: let the observer $\mathcal{O}'$ move with velocity v relative to $\mathcal{O}$ in z-direction and consider a light ray in the z–x-plane, which enters the eye of $\mathcal{O}$ at $t = 0$ with an angle θ to the z-axis. In his coordinates the light traverses the worldline Γ: $t \mapsto l(t) = t(1, -\mathbf{e}) , \ \mathbf{e} = (\sin\theta, 0, \cos\theta)$ (2.36).[1] For $\mathcal{O}'$ the same light ray enters with some angle θ' to the opposite of the direction into which he sees $\mathcal{O}$ move. $\mathcal{O}'$ attributes the worldline $t' \mapsto t'(1, -\mathbf{e}') , \ \mathbf{e}' = (\sin\theta', 0, \cos\theta') = \Lambda l(t)$ to the same light ray.

The rotation-free Lorentz boost in z-direction leaves the x- and y-coordinates invariant and stretches the difference $t_- = t - z$ by κ (2.10), $t_-' = \kappa t_-$ (3.2). Consequently for the observer $\mathcal{O}'$ the ratio $l_y/l_- = \sin\theta/(1 + \cos\theta)$ is diminished by a factor κ.

---

[1] The light pulse moves into the opposite of the direction **e** from which it is seen.

Fig. 3.3 Circle with $\sin\theta$ and $1+\cos\theta$

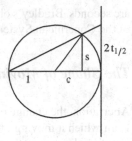

Using the trigonometric identity (Fig. 3.3)

$$\frac{\sin\theta}{1+\cos\theta} = \frac{2\cos\frac{\theta}{2}\sin\frac{\theta}{2}}{2\cos^2\frac{\theta}{2}} = \tan\frac{\theta}{2} \qquad (3.18)$$

the angle θ' of the incident light ray, seen by the observer $\mathcal{O}'$, is related by

$$\tan\frac{\theta'}{2} = \sqrt{\frac{1-v}{1+v}}\,\tan\frac{\theta}{2}\,. \qquad (3.19)$$

to the angle θ which $\mathcal{O}$ observes [29].

The change of the direction of the incident light ray is called aberration.

The tangent increases monotonously with the angle, therefore θ' is smaller than θ for $0 < v < 1$. Just as rain incident light comes more from the direction into which an observer moves.

Yearly Aberration of the Light of Stars

The earth orbits the sun at a mean radius of $r = 1.50\cdot10^{11}$ m. The length $2\pi r$, covered within a year, as compared to a light year, $9.46\cdot10^{15}$ m, is the velocity $v = 1.00\cdot10^{-4}$. For such small velocities $\delta\theta = \theta' - \theta$ is small and (3.19) is approximately

$$\tan\frac{\theta}{2} + \delta\theta\,\frac{1}{2\cos^2(\theta/2)} \approx \tan\frac{\theta'}{2} = \sqrt{\frac{1-v}{1+v}}\,\tan\frac{\theta}{2} \approx (1-v)\tan\frac{\theta}{2}\,,\ \delta\theta \approx -v\sin\theta\,. \qquad (3.20)$$

A star which for a stationary observer lies in a direction $\theta = \frac{\pi}{2}$ orthogonal to the current direction of motion of the earth is seen from an observer on earth in a direction shifted by $|\delta\theta| = 10^{-4}$, i.e. 20.5 arc seconds.[2]

During each year the direction of the velocity of the terrestrial observers changes. Therefore, as discovered by James Bradley in 1728, we perceive distant stars in directions which traverse in the course of the year ellipses with a major axis of 41

---

[2] As meter per second (1.3), an arc second is just a number, $1'' = 2\pi/(360\cdot60\cdot60) \approx 4.848\cdot10^{-6}$.

arc seconds. Bradley's observation of the aberration of stars was the first direct proof of the Copernican System that the earth orbits the sun.

The Shape of Moving Spheres

Aberration, the change of directions $\mathbf{e}_{\theta,\varphi}$ of incident light rays to the directions $\mathbf{e}_{\theta'\varphi'}$ from which a moving observer sees the incoming light, is an invertible map of the set of all directions, the two-dimensional sphere S^2, to itself.

Moving observers do not see spheres length-contracted to pan cakes but again as spheres [18, 29, 33, 36]. So aberration is a map of S^2 to itself which maps circles, the outline of spheres, to circles. This is concluded from the following arguments. All directions $\mathbf{e}$, $\mathbf{e}^2 = 1$ of incident light rays from the outline of a sphere constitute a circular cone with some opening angle δ to the axis $\mathbf{n}$, $\mathbf{n}^2 = 1$,

$$\cos \delta = \mathbf{e} \cdot \mathbf{n}. \tag{3.21}$$

The light ray which an observer at the origin sees from the direction $\mathbf{e}$ traverses the worldline

$$\Gamma : t \mapsto l(t) = tk, \quad k = (1, -\mathbf{e}), \quad k^2 = 0. \tag{3.22}$$

Its tangent vector, k, is lightlike and belongs to the circular cone (3.21) if and only if k is orthogonal (in the sense of the scalar product (2.45)) to the spacelike four-vector

$$n = a(-\cos \delta, \mathbf{n}), \quad n^2 < 0, \tag{3.23}$$

$$k \cdot n = a(-\cos \delta + \mathbf{e} \cdot \mathbf{n}) = 0. \tag{3.24}$$

Since Lorentz transformations preserve scalar products (2.51), the transformed vectors $n' = \Lambda n$ and the transformed tangent vectors $k' = \Lambda k$ satisfy the equation $k'^2 = 0, n'^2 < 0$ and $k' \cdot n' = 0$. Because each spacelike vector is of the form (3.23), the vector $n' = \Lambda n$ defines an axis $\mathbf{n}'$ and an opening angle δ' of a circular cone of light rays. For a moving observer the light rays transformed by aberration come from directions which form a circular cone.

Infinitesimal cones with opening angle δ are scaled by a factor D which is obtained by differentiation of $\theta'(\theta)$ (3.19)

$$D = \frac{d\theta'}{d\theta} = \frac{1}{\kappa} \frac{\cos^2(\theta'/2)}{\cos^2(\theta/2)} = \frac{1}{\kappa(1 + \kappa^{-2} \tan^2(\theta/2)) \cos^2(\theta/2)} = \frac{\sqrt{1 - v^2}}{1 + v \cos \theta}. \tag{3.25}$$

Because a small circle is magnified by a factor D, each of its diameters appear enlarged by this factor irrespective of its direction. Moreover the scale factor D of nearby objects is nearly the same.

Pictures, which are perceived by observers in relative motion at the same place in the same instant, agree in the relative sizes of nearby, small objects. In particular, angles between intersecting lines are preserved by aberration because they are the ratio of small circular arcs to small radii of circles: aberration is conformal.

That aberration is conformal also follows simply from the fact that by the arguments before (3.18) it scales the stereographic projection

$$(x, y, z)|_{x^2+y^2+z^2=1} \mapsto (u, v) = \left(\frac{x}{1-z}, \frac{y}{1-z} \right) \tag{3.26}$$

of the sphere $x^2 + y^2 + z^2 = 1$ of the directions $\mathbf{e} = (x, y, z)$ of incident light rays by $1/\kappa$ (3.19). This dilation is a conformal map and also the projection (3.26) is conformal, because on the unit sphere one has

$$1-z = \frac{1}{2}\left(x^2+y^2+(z-1)^2\right), \quad u^2+v^2+1 = \frac{x^2 + y^2 + (z-1)^2}{(1-z)^2} = \frac{2}{1-z}. \tag{3.27}$$

If therefore, (x, y, z) lie on a circle on S^2 and satisfy $n_x x + n_y y + n_z z = \cos \delta$, then dividing this linear equation by $(1-z)$ one obtains

$$n_x u + n_y v + \frac{n_z}{1-z} = n_z + \cos \delta \,, \tag{3.28}$$

and because of $2/(1-z) = u^2 + v^2 + 1$ this is a quadratic equation for u and v with the same coefficient for u^2 and v^2, i.e. a circle in the u–v-plane.

Moving Ruler

Let us investigate, as illustrated in Fig. 3.4, the one-dimensional image of the edges of a flat infinitesimal ruler in the x–y-plane with sides dx and dy parallel to the axes. An observer at rest looking at the ruler under an angle $\theta = \alpha + \pi/2$ to the x-axis from a distance $a = A/\cos\alpha$ perceives the radians $dx \cos\alpha/a$ and $dy \sin\alpha/a$.

At the same time and at the same position an observer moving in x-direction with velocity v sees the edges of a ruler flying past him where all pixels are shifted by aberration as compared to the stationary observer and form an angle $\theta' = \alpha' + \pi/2$ with the x-axis. Since aberration is conformal, the length ratio of the visible edges of the ruler is the same for the moving observer and the stationary observer. Both observers therefore see the projection of a ruler rotated by α perpendicular to the line of sight. If one observes a moving ruler with an angle θ' to the x-axis and compares it with a ruler at rest, it appears to be rotated by $\theta'(v, \theta) - \theta$.

Because aberration preserves the relative size of small, nearby objects, a moving ruler does not appear contracted in any direction but is seen scaled by $\sin\theta'/\sin\theta = \cos\alpha'/\cos\alpha$ (3.25). The moving observer therefore sees in the direction α' the edges of a ruler at the distance $A/\cos\alpha'$. This is the distance a ruler at rest has on the x-axis shifted by A, if it is seen under the angle α'. The visible size of the moving ruler

Fig. 3.4 Moving ruler

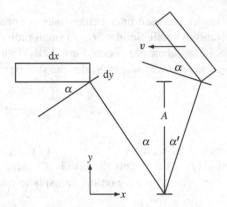

does not depend on its velocity but only on the retarded position which it had for the observer when it emitted its light.

An object at rest is observed as it is and where it is, even if one perceives its light later. A moving object is not seen where it is but *where it was* with the corresponding size. Each aspect ratio of visible parts, however, is seen *as it is* perceived by a co-moving observer at the same instant at the same place.

A concealed side of the ruler cannot be seen before the ruler has passed the observer. For the stationary observer the right, short edge of the ruler is visible only if it is behind him in x-direction. Then it is also behind the moving observer, for if two mutually moving observers meet at some time and some position they agree upon which events occur at this time behind or ahead of them in direction of the relative motion. In particular, the events $(t = 0, x = 0, y, z)$ which separate behind and ahead are invariant under rotation-free Lorentz transformations in x-direction.

A light ray which is emitted with an angle $\bar{\theta}$ with

$$\cos \bar{\theta} = v \qquad (3.29)$$

to the direction of motion **v** of a particle reaches the observer in the same instant as the particle is on his side. If seen under this angle a uniformly moving particle is next to the observer. At this moment only the sides of the particle are visible; the co-moving observer, for whom the particle is at rest, observes it under an angle of $\frac{\pi}{2}$ besides, above or below himself. Because this angle separates what is behind and what is before the observer, we call it Janus angle after the Roman god of doors and doorways, just as January lies between the old year and the remainder of the new year (Fig. 3.5).

Fig. 3.5 Janus angle

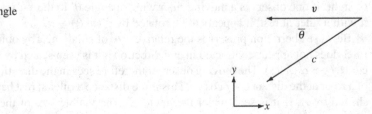

Luminosity

The images, which moving observers perceive, are not only deformed but also changed in their luminosity and color. We denote with $n(\omega, \theta, \varphi)\mathrm{d}\omega \, \mathrm{d}t \, \mathrm{d}\Omega$ the number of photons in the frequency interval $\mathrm{d}\omega$ which an observer perceives in the time $\mathrm{d}t$ in the direction (θ, φ) in the solid angle segment $\mathrm{d}\Omega = \sin\theta \mathrm{d}\theta \, \mathrm{d}\varphi$. The observer moving in direction $\theta = 0$ perceives photons in the Doppler-shifted frequency interval $\mathrm{d}\omega' = D^{-1}\mathrm{d}\omega$ (2.30) in the solid angle segment $\mathrm{d}\Omega' = D^2\mathrm{d}\Omega$ modified by aberration (3.25). The time interval $\mathrm{d}t'$ in which the moving observer at the same position and the same moment in time sees the same number of photons is given by $\mathrm{d}t' = D\mathrm{d}t$. This one infers most easily from the fact that for a uniform photon current the number of photons per time defines a frequency just as the number of vibrations of the wave per time. Because both observers detect the same number of photons,

$$n' \, \mathrm{d}\omega' \, \mathrm{d}t' \, \mathrm{d}\Omega' = n' \, D^2 \, \mathrm{d}\omega \, \mathrm{d}t \, \mathrm{d}\Omega = n \, \mathrm{d}\omega \, \mathrm{d}t \, \mathrm{d}\Omega \, , \tag{3.30}$$

and the moving observer sees the spectral photon current density

$$n'(\omega', \theta', \varphi') = \frac{(1 + v \, \cos\theta)^2}{1 - v^2} n(\omega, \theta, \varphi) \, . \tag{3.31}$$

3.3 Energy and Momentum

A conserved quantity is a function $\phi(t, \mathbf{x}, \mathbf{v})$ of the time t, the position $\mathbf{x}$ and the velocity $\mathbf{v}$ which on the paths of physical particles retains its initial value,

$$\phi\left(t, \mathbf{x}(t), \frac{\mathrm{d}\mathbf{x}}{\mathrm{d}t}(t)\right) = \phi\left(0, \mathbf{x}(0), \frac{\mathrm{d}\mathbf{x}}{\mathrm{d}t}(0)\right) . \tag{3.32}$$

For example, in Newtonian physics the energy E and the momentum $\mathbf{p}$ of a free particle with mass m

$$E = E_0 + \frac{1}{2}m\mathbf{v}^2, \quad \mathbf{p} = m\mathbf{v} , \tag{3.33}$$

are conserved due to the equation of motion

$$\frac{\mathrm{d}\mathbf{p}}{\mathrm{d}t} = 0, \quad \frac{\mathrm{d}\mathbf{x}}{\mathrm{d}t} = \frac{1}{m}\mathbf{p} . \tag{3.34}$$

The value of the energy for a particle at rest is irrelevant in Newtonian physics, usually E_0 is simply set to zero.

The mass m is trivially conserved during the motion, because it is a constant which depends neither on the position nor on the velocity. It is *not* conserved in decays (3.53).

Transformation of Additive Conserved Quantities

Of course, for a free particle all functions of the velocity are conserved quantities since the velocity is constant for free motion. The outstanding importance of energy and momentum comes from the fact that they are additive conserved quantities, i.e. the sums of the momenta and the energies of several particles are conserved quantities even if the individual momenta and energies change through, say, elastic collisions.

If an observer detects conserved quantities ϕ, then there are conserved quantities also for each other uniformly moving observer with coordinates $x' = \Lambda x + a$ (3.10), and there is a transformation which allows to convert the conserved quantities of both observers.

Additive conserved quantities have to transform linearly

$$(\phi_{(1)} + \phi_{(2)})' = \phi'_{(1)} + \phi'_{(2)} , \quad (c\phi)' = c\phi' , \tag{3.35}$$

because they are sums and multiples of parts for both observers. The transformation therefore is similar to Lorentz transformations and of the form

$$\phi' = M_{\Lambda,a}\phi . \tag{3.36}$$

The matrices $M_{\Lambda,a}$ are restricted by the fact that successive transformations can be evaluated in steps or directly (3.11) also for the transformations of additive conserved quantities,

$$\phi'' = M_{\Lambda_2 \circ 1,\, a_{2\circ 1}}\phi = M_{\Lambda_2,\, a_2} M_{\Lambda_1,\, a_1}\phi , \tag{3.37}$$

which has to hold for arbitrary values of ϕ. Therefore the product of matrices $M_{\Lambda,a}$ has to give the matrix which corresponds to the successively applied transformations

$$M_{\Lambda_2 \Lambda_1,\, a_2 + \Lambda_2 a_1} = M_{\Lambda_2,\, a_2}\, M_{\Lambda_1,\, a_1} . \tag{3.38}$$

Matrices M_g, which correspond to the elements g of a group G such that their products $M_{g_2} M_{g_1} = M_{g_2 \circ g_1}$ correspond to the products $g_2 \circ g_1$ of the group elements, constitute a representation of the group G. The question, which representations there are for a given group, has been studied mathematically at length.

Four-Momentum

In the simplest example four conserved quantities $p = (p^0, p^1, p^2, p^3)$, which in anticipation of the result of our investigation we shall call four-momentum, transform according to

$$p' = \Lambda p . \tag{3.39}$$

Under translations $x' = x + a$, i.e. $\Lambda = \mathbf{1}$, which shift time and position by $a = (a^0, a^1, a^2, a^3)$ the four-momentum is unchanged. Thus, it does not depend on the position or the time but only on the velocity of the particle.

If a particle is slower than light, then there is a reference frame of a co-moving observer for whom the particle is at rest. Since the velocity $\mathbf{v} = 0$ is invariant under rotations and since the four-momentum is a function of the velocity, rotations do not change the four-momentum p of a particle at rest. Accordingly, in the rest frame of a particle the spatial part $\mathbf{p} = (p^1, p^2, p^3)$ of the four-momentum vanishes and it has the form

$$p_{\text{rest}} = (m, 0, 0, 0) . \tag{3.40}$$

If we transform the four-momentum with (3.8) into the reference frame of an observer for whom the particle moves with velocity v in x-direction

$$\begin{pmatrix} p^0 \\ p^1 \end{pmatrix} = \frac{1}{\sqrt{1 - v^2}} \begin{pmatrix} 1 & v \\ v & 1 \end{pmatrix} \begin{pmatrix} m \\ 0 \end{pmatrix} = \begin{pmatrix} \frac{m}{\sqrt{1-v^2}} \\ \frac{m v}{\sqrt{1-v^2}} \end{pmatrix}, \tag{3.41}$$

and if we rotate the motion into an arbitrary direction, we obtain

$$p^0 = \frac{m}{\sqrt{1 - \mathbf{v}^2}} , \quad \mathbf{p} = \frac{m \mathbf{v}}{\sqrt{1 - \mathbf{v}^2}} . \tag{3.42}$$

We denote the components of the four-momentum with the same names as the quantities in Newtonian physics with which they coincide in the limit of small velocities. Up to higher powers of $\mathbf{v}$ one has

$$p^0(\mathbf{v}) = m + \frac{1}{2}m\mathbf{v}^2 + \cdots , \quad \mathbf{p}(\mathbf{v}) = m \mathbf{v} + \cdots . \tag{3.43}$$

Therefore (making factors c explicit for once) $E = cp^0$ is the energy

$$E(\mathbf{v}) = \frac{mc^2}{\sqrt{1 - \frac{\mathbf{v}^2}{c^2}}} , \tag{3.44}$$

$\mathbf{p}$ is the momentum and m is the mass of the particle. It is positive, and the energy is bounded from below.

For a particle moving with the speed of light, we choose (in units with $c = 1$) a reference frame in which $\frac{dx}{ds}$ points into x-direction. The tangential vector to the world line of the particle then is of the form $\frac{dx}{ds} = \frac{dt}{ds}(1, 1, 0, 0)$. It is invariant under rotations about the x-axis and invariant under multiplication with

$$\Lambda_a = \exp \begin{pmatrix} 0 & 0 & -a & 0 \\ 0 & 0 & -a & 0 \\ -a & a & 0 & 0 \\ 0 & 0 & 0 & 0 \end{pmatrix} = \begin{pmatrix} 1+\frac{a^2}{2} & -\frac{a^2}{2} & -a & \\ \frac{a^2}{2} & 1-\frac{a^2}{2} & -a & \\ -a & a & 1 & \\ & & & 1 \end{pmatrix}. \tag{3.45}$$

These are Lorentz matrices because the column vectors have length squared ± 1 and are mutually orthogonal.[3]

The four-momentum of the particle, moving with the velocity of light, must also be invariant under rotations around the x-axis and invariant under multiplication with the matrix (3.45), since it is a function of $\frac{dx}{ds}$. Thus, the components p^2 and p^3 must vanish and $p^0 = p^1$ has to hold. If the particle is moving in an arbitrary direction, one has

$$p^0 = |\mathbf{p}|, \quad E = |\mathbf{p}|. \tag{3.46}$$

Also for particles which move with the velocity of light the four-momentum is a multiple of the tangent vector of the world line. For massive and for massless particles the velocity $\mathbf{v}$ is the ratio

$$\mathbf{v} = \frac{\mathbf{p}}{p^0} = \frac{\mathbf{p}}{\sqrt{m^2 + \mathbf{p}^2}}. \tag{3.47}$$

The velocity is a function of the momentum and the momentum is conserved. This is the statement that particles are inert. To change their velocity forces have to transfer momentum.

The vacuum is the same for all observers and therefore has a four-momentum which is invariant under all transformations $p' = \Lambda p$. Therefore its energy and its momentum must vanish $p_{\text{vacuum}} = (0, 0, 0, 0)$. This also holds for the contribution of the so-called quantum fluctuations to the energy which trouble some theoretical physicists.

Mass

The mass is the length of the four-momentum of a free particle. No matter what its velocity is, in units with $c = 1$ it satisfies

---

[3] The matrices Λ_a arise from repeated infinitesimal transformation $\omega = d\Lambda_a/da)_{|a=0}$ where the series $\Lambda_a = \exp a\omega = 1 + a\omega + a^2\omega^2/2$ consists of three terms only.

$$p^2 = (p^0)^2 - \mathbf{p}^2 = m^2 \,, \quad p^0 = \sqrt{m^2 + \mathbf{p}^2} > 0 \,. \tag{3.48}$$

This is the equation for a sheet of a hyperboloid: the four-momenta of a free particle lie on the mass-shell.

The relation (3.48) of energy and momentum also holds for particles which move with the velocity of light, for instance for photons. They are massless. Their four-momentum p is lightlike

$$p^2 = (p^0)^2 - \mathbf{p}^2 = 0 \,, \quad p^0 = |\mathbf{p}| > 0 \,. \tag{3.49}$$

Photons with four-momentum p correspond to quanta of plane electromagnetic waves (5.81) with four-wave-vector $k = (\omega = |\mathbf{k}|, \mathbf{k})$. Here the momentum $\mathbf{p} = \hbar\mathbf{k}$ is a multiple of the wave-vector, the factor $\hbar = 1.055 \cdot 10^{-34}$ Js [28] is Planck's constant. The energy $E = \hbar\omega = \hbar|\mathbf{k}|$ of the photons is a multiple of the frequency $\nu = \omega/(2\pi)$ of the electromagnetic wave. This relation is fundamental for Planck's derivation of the thermal radiation density and Einstein's interpretation of the photoelectric effect.

According to (3.44) particles at rest have the energy

$$E_{\text{rest}} = mc^2 \,. \tag{3.50}$$

This is probably the most famous equation in physics. It underlies the realization that during the transmutation of atomic nuclei by fission or fusion energy can be released, for the total mass of the nuclei is measurably different from the sum of the individual masses. The mass difference is due to the binding energy which can be used in war or peace, destructively or beneficially. Equation (3.44) also contains the statement that it requires infinite energy to accelerate a massive particle to the speed of light. Massive particles are always slower than light.

The mass m is independent of the velocity. In some presentations of the theory of relativity the term "mass" is used for the product γm with the velocity dependent factor $\gamma = 1/\sqrt{1 - v^2}$. This is a waste of a denomination, because for $E = \gamma mc^2$ we already have a name: energy. These days one denotes with mass the quantity which in old and outdated presentations goes by the long-winded "rest mass".

Moreover, if one denotes γm "mass" it is tempting to insert this into formulas of Newtonian physics, which emerge in the limit of low velocities from relativistic physics, expecting to obtain equations, which are valid for all velocities. Even if this is true in one case for the momentum $\mathbf{p} = \gamma m\mathbf{v}$ one nearly always obtains nonsense: the kinetic energy is neither $\gamma m\mathbf{v}^2/2$ nor $\mathbf{p}^2/(2\gamma m)$.

A particle does not become heavier with increasing speed: weight depends on the acceleration as compared to free fall. A fast moving particle does not generate the gravity of a mass, magnified by a factor γ. It does not become a black hole by rapid motion. If it required only a factor γ, then Einstein would have needed ten minutes rather than ten years to include gravity into his relativistic formulation of mechanics and electrodynamics.

Force is not mass times acceleration. The equations of motion of a relativistic charged particle state, as we will show in Sect. 5.2, that the total momentum of the particle and the fields remains conserved,

$$\mathbf{F} = \frac{d\mathbf{p}}{dt} . \tag{3.51}$$

The force $\mathbf{F}$ is the momentum per time $d\mathbf{p}/dt$ transferred from the electromagnetic fields to the particle.

The particle is generally not accelerated into the direction of the force

$$\frac{d\mathbf{v}}{dt} = \sum_i \frac{\partial \mathbf{v}}{\partial p^i} \frac{dp^i}{dt} \overset{3.47}{=} \frac{\mathbf{F}}{\sqrt{m^2 + \mathbf{p}^2}} - \frac{(\mathbf{p} \cdot \mathbf{F})\mathbf{p}}{\sqrt{m^2 + \mathbf{p}^2}^3} = \frac{1}{\sqrt{m^2 + \mathbf{p}^2}}(\mathbf{F} - (\mathbf{v} \cdot \mathbf{F})\mathbf{v}) . \tag{3.52}$$

Inertia of fast particles depends on the direction of the velocity $\mathbf{v}$. A transverse force causes an acceleration $d\mathbf{v}_\perp/dt = \sqrt{1 - v^2}\mathbf{F}_\perp/m$; parallel to the velocity, the particle is more inert by a factor $1/(1 - v^2)$. Also massless particles are inert, $d\mathbf{v}_\perp/dt = \mathbf{F}_\perp/|\mathbf{p}|$, in their direction of motion even infinitely inert, $d\mathbf{v}_\parallel/dt = 0$.

If one defined force to denote mass times acceleration, it would not satisfy "actio et reactio" because momentum, not mass times velocity, is conserved.

If macroscopic bodies move and interact with each other, then effects from the finite speed of sound become large long before relativistic effects are measurable. At high relative velocities the binding of the constituents of macroscopic bodies becomes negligible, they do not behave as rigid bodies but nearly as a gas of free particles which collide, restricted by energy and momentum conservation, and which interact with fields like the electromagnetic or the gravitational field.

Decay into Two Particles

If a particle at rest with mass m decays into two particles with masses m_1 and m_2, then the energies of the decay products are fixed by the masses involved. Due to momentum conservation the momentum $\mathbf{p}$ of the first decay product is the opposite of the momentum of the second particle. The energies of the two decay products are $E_1 = \sqrt{m_1^2 + \mathbf{p}^2}$ and $E_2 = \sqrt{m_2^2 + \mathbf{p}^2}$, since energy and momentum lie on the mass shell (3.48). Energy conservation implies that the sum of these energies coincides with the energy m of the decaying particle at rest

$$m = \sqrt{m_1^2 + \mathbf{p}^2} + \sqrt{m_2^2 + \mathbf{p}^2} > m_1 + m_2 . \tag{3.53}$$

In particular, the mass m of the decaying particle is not conserved. It is greater than the sum of the masses of the decay products. This is the same geometric fact as with the twin paradox that the sum $p_1 + p_2$ of two timelike four-vectors is longer than the sum of the individual lengths of p_1 and p_2.

Repeatedly taking the square and reordering terms one solves for $\mathbf{p}$ and E_1

$$\mathbf{p}^2 = \frac{1}{4m^2}(m^4 + m_1^4 + m_2^4 - 2m^2 m_1^2 - 2m^2 m_2^2 - 2m_1^2 m_2^2), \qquad (3.54)$$

$$E_1 = \frac{1}{2m}(m^2 - m_2^2 + m_1^2). \qquad (3.55)$$

If tachyons existed with a spacelike four-momentum p, $p^2 = -m^2$, then the modulus of its momentum, $|\mathbf{p}| = \sqrt{m^2 + E^2}$, not its energy E, would be bounded from below. If they interacted electromagnetically they could emit photons with arbitrary large energy, because the tachyon mass shell is a doubly-ruled surface [35] and the four-momentum of the tachyon after the emission of the photon with four-momentum k, $k^2 = 0$, is again on-shell, $(p - k)^2 = -m^2$, if $p \cdot k = 0$, i.e. (3.24) if the photon is emitted with an angle $\cos \theta = p^0/|\mathbf{p}|$ to the momentum of the tachyon.

Compton Scattering

On elastic scattering of two particles, i.e. for a scattering process where the number of particles and their masses remain unchanged, the conservation of energy and momentum fixes the energies after the collision as a function of the scattering angle θ and the initial energies.

Let us consider, for example, an incident photon with energy E which is scattered elastically from an electron initially at rest. This process is called Compton scattering.

Let p and q be the four-momenta of the photon and the electron before and p' and q' after the scattering. Four-momentum conservation implies

$$p + q = p' + q', \qquad (3.56)$$

or in more detail, if we choose the x-axis in the initial direction of motion of the photon and the y-axis such that the photon, scattered by an angle θ, finally moves in the x–y-plane,

$$\begin{pmatrix} E \\ E \\ 0 \\ 0 \end{pmatrix} + \begin{pmatrix} m \\ 0 \\ 0 \\ 0 \end{pmatrix} - \begin{pmatrix} E' \\ E' \cos \theta \\ E' \sin \theta \\ 0 \end{pmatrix} \stackrel{!}{=} \begin{pmatrix} m + E - E' \\ E - E' \cos \theta \\ -E' \sin \theta \\ 0 \end{pmatrix}, \qquad (3.57)$$

Here we have already taken into account that the scattered photon with energy E' is massless and satisfies $p'^2 = 0$. That also the electron is on its mass shell after the collision, $q'^2 = m^2$, relates the scattering angle and the energies

$$(m + E - E')^2 - (E - E' \cos \theta)^2 - (E' \sin \theta)^2 = m^2. \qquad (3.58)$$

The squares of E, E' and m cancel. With the conventional factors c one obtains

$$\frac{mc^2}{E'} = \frac{mc^2}{E} + 1 - \cos \theta. \qquad (3.59)$$

The energy of the outgoing photon therefore is fixed by the scattering angle. The energy is smaller than the energy of the incoming photon. This contradicts the notion of an incoming electromagnetic wave corresponding to the photon of energy $E = \hbar \omega$ which accelerates the charged electron, which then itself emits a wave with the scattered photons. In such a process the frequency of the emitted wave would coincide with the original frequency. Equation (3.59) on the other hand follows from the assumption that electrons are particles and that electromagnetic waves consist of particles, namely photons.

However, this is not a proof of the particle property of electromagnetic waves. One may also obtain (3.59) if one—which we did not do—treats the electron as well as the photon as a wave. That waves behave like particles and that particles have properties of waves belongs to the basic facts of quantum physics.

Chapter 4
Relativistic Particles

Abstract The time evolution of particles and fields is characterized by a correspond-ing functional, the action. It is stationary for the real evolution as compared to other imaginable evolutions. This principle of the stationary action allows to understand the conservation of energy, momentum and angular momentum as consequences of symmetries, because to each infinitesimal symmetry of the action there corresponds a conserved quantity and, vice versa, to each conserved quantity there corresponds an infinitesimal symmetry of the action. Physical and geometrical properties, con-servation laws and symmetries, are intimately related by this Noether theorem.

4.1 Clocks on Worldlines

In the course of time a clock traverses its worldline f, parameterized by some para-meter s from some real interval I,

$$f : \begin{cases} I \subset \mathbb{R} \to \mathbb{R}^4 \\ s \quad \mapsto f(s) \end{cases} \tag{4.1}$$

and shows in each event $f(s) = (f^0(s), f^1(s), f^2(s), f^3(s))$ the time $\tau(s)$ which goes by on the clock. The coordinate time f^0 is assumed to increase monotonically. Then the events on the worldline are traversed precisely once and causally ordered.

On a straight worldline the time on the clock increases by (2.43)

$$\Delta \tau = ds \sqrt{\left(\frac{df^0}{ds}\right)^2 - \left(\frac{df^1}{ds}\right)^2 - \left(\frac{df^2}{ds}\right)^2 - \left(\frac{df^3}{ds}\right)^2} = ds \sqrt{\frac{df}{ds} \cdot \frac{df}{ds}} \tag{4.2}$$

between adjacent events with difference vector $ds \left(\frac{df^0}{ds}, \frac{df^1}{ds}, \frac{df^2}{ds}, \frac{df^3}{ds}\right)$. The root is real, because events on worldlines of clocks are mutually timelike and the tangent vector $\frac{df}{ds}$ has positive length squared.

N. Dragon, *The Geometry of Special Relativity—a Concise Course,*
SpringerBriefs in Physics, DOI: 10.1007/978-3-642-28329-1_4,
© The Author(s) 2012

If the worldline f is not straight but accelerated, it is approximated by an increasingly finer sequence of small straight segments. An ideal clock measures and adds the times which pass on these segments.

Time: *An ideal clock records the time*

$$\tau[f] = \int_f \Delta\tau = \int_{\underline{s}}^{\overline{s}} \mathrm{d}s \sqrt{\left(\frac{\mathrm{d}f^0}{\mathrm{d}s}\right)^2 - \left(\frac{\mathrm{d}f^1}{\mathrm{d}s}\right)^2 - \left(\frac{\mathrm{d}f^2}{\mathrm{d}s}\right)^2 - \left(\frac{\mathrm{d}f^3}{\mathrm{d}s}\right)^2} \qquad (4.3)$$

between the events $A = f(\underline{s})$ and $B = f(\overline{s})$ on its worldline f.

In contrast to Newtonian physics the time on the clock is not a function of the events A and B, and $\Delta\tau$ is not the derivative $\mathrm{d}\tau$ of a function τ, which is defined in spacetime. The time on the clock does not pertain to the events A and B alone but is the path length of the worldline f from A to B.

Time is additive. If C is an event on the worldline f in between A and B and if f_1 and f_2 denote the parts of the worldline to and from C, then in a suggestive notation $f = f_1 + f_2$ and

$$\int_{f_1+f_2} \Delta\tau = \int_{f_1} \Delta\tau + \int_{f_2} \Delta\tau. \qquad (4.4)$$

The time $\tau[f]$ is independent of the parameterization of the worldline.

This holds because each other parameterization $f' : s' \mapsto f'(s')$ of the worldline with monotonically increasing coordinate time f'^0 is given by $f'(s') = f(s(s'))$ with monotonically increasing $s(s')$, therefore $\frac{\mathrm{d}s}{\mathrm{d}s'} = \sqrt{\left(\frac{\mathrm{d}s}{\mathrm{d}s'}\right)^2}$. By the chain rule one has $\frac{\mathrm{d}f'}{\mathrm{d}s'} = \frac{\mathrm{d}s}{\mathrm{d}s'}\frac{\mathrm{d}f}{\mathrm{d}s}$, and with the integral substitution theorem the assertion follows,

$$\int_{\underline{s}'}^{\overline{s}'} \mathrm{d}s' \sqrt{\frac{\mathrm{d}f'}{\mathrm{d}s'} \cdot \frac{\mathrm{d}f'}{\mathrm{d}s'}} = \int_{\underline{s}'}^{\overline{s}'} \mathrm{d}s' \frac{\mathrm{d}s}{\mathrm{d}s'} \sqrt{\frac{\mathrm{d}f}{\mathrm{d}s} \cdot \frac{\mathrm{d}f}{\mathrm{d}s}} = \int_{\underline{s}}^{\overline{s}} \mathrm{d}s \sqrt{\frac{\mathrm{d}f}{\mathrm{d}s} \cdot \frac{\mathrm{d}f}{\mathrm{d}s}}. \qquad (4.5)$$

If one parameterizes the worldline with the coordinate time, $f^0(s) = s$, then the worldline is given by $f : t \to (t, f^1(t), f^2(t), f^3(t))$ and the tangent vector has the components $\frac{\mathrm{d}f}{\mathrm{d}t} = (1, \frac{\mathrm{d}f^1}{\mathrm{d}t}, \frac{\mathrm{d}f^2}{\mathrm{d}t}, \frac{\mathrm{d}f^3}{\mathrm{d}t}) = (1, v_x, v_y, v_z)$ and the length squared $1 - \mathbf{v}^2$. So the clock registers the time

$$\tau[f] = \int_{\underline{t}}^{\overline{t}} \mathrm{d}t \sqrt{1 - \mathbf{v}^2(t)}. \qquad (4.6)$$

If one chooses the parameter s along the worldline such that the tangent vector has unit length $\left(\frac{\mathrm{d}f}{\mathrm{d}s}\right)^2 = 1$ everywhere, then s coincides up to a constant with τ, the proper time which passes on the clock. Conversely, the tangent vector has unit length everywhere if the worldline is parameterized with its proper time,

$$\left(\frac{\mathrm{d}f}{\mathrm{d}s}\right)^2 = 1 \iff \tau[f] = \bar{s} - \underline{s} \,. \tag{4.7}$$

On straight worldlines τ^2 coincides with the length squared (2.46) of the difference vector w_{BA} (2.52) from A to B.

The proper time $\tau[f]$ is independent of the acceleration in the sense that the integrand $\Delta\tau$ (4.2) does not depend on the second derivatives $\mathrm{d}^2 f/\mathrm{d}s^2$. Yet, the time $\tau[f]$ depends on the worldline f and has different values for straight and for curved worldlines from A to B, even though the clock is independent of acceleration.

By definition an ideal clock records the path length of the worldline irrespective of its acceleration. Whether a real clock is ideal has to be checked by measurement.

Many real clocks deviate considerably from ideal clocks. Sun dials display the angle of an earthbound axis to the sun. Pendulum clocks and hour glasses measure the acceleration. Paradoxically, an hour glass runs if one keeps hold of it and stops if one let it loose and drops it, even before it hits the ground. Each pendulum circulates with constant angular velocity if one drops the pendulum clock. Quartz clocks change their frequency if the accelerating forces deform the quartz crystal. In quantum mechanical systems the modification of the Hamiltonian, which causes the acceleration, normally also changes the energy levels with which the clock operates. If, for example, atoms are stored in a magnetic field it detunes the transition frequencies.

Clocks can be realized by completely different physical processes, for instance by comparison with the rotation of the earth or by electromagnetic transitions in atoms. After the exclusion of obviously unsuitable clocks, e.g. moving sun dials, and after correction of the known imperfections of real clocks, in particular after gravitational corrections, (4.3) agrees without exception with all observations. Among others, it underlies the everyday operation of the global positioning system [27, 30].

To check our understanding of elementary particles physicists [3] keep muons in a cyclotron on a circular path and compare the observed rotation of the muon spin in the magnetic field to the theoretical predictions with a precision of nine decimals. All observations are compatible with the simple assumption that the inner clock of the muons, which controls their decay, records the time (4.3) independently of the acceleration. There was no deviation from (4.3) within the measurement precision of 1 %, even though for high energy muons the acceleration in the cyclotron, a, amounts to roughly $4 \cdot 10^{16}$ times the acceleration by the gravity of earth. However, in terms of the muon mass $m_\mu c^2 = 105\,\mathrm{MeV}$ [28], Planck's constant $\hbar = 1{,}05 \cdot 10^{-34}\,\mathrm{J\,s}$ and the speed of light c, the acceleration is still small, $a\hbar/(m_\mu c^3) \approx 10^{-13}$. This number is the change of the velocity of the muon in the cyclotron relative to the speed of light during one oscillation of its wave function.

4.2 Free Particles

On straight worldlines, the worldlines of free particles, more times passes between each pair of events A and B than on each other worldline from A to B.

To show this one considers a worldline $s \mapsto f(s)$ and evaluates the time τ for neighboring curves $s \mapsto f(s) + \delta f(s)$ which also pass through A and B, $\delta f(\underline{s}) = 0$ and $\delta f(\overline{s}) = 0$. Then τ changes to first order in δf by

$$
\delta\tau = \int\limits_{\underline{s}}^{\overline{s}} ds \left(\sqrt{\frac{d(f+\delta f)}{ds} \cdot \frac{d(f+\delta f)}{ds}} - \sqrt{\frac{df}{ds} \cdot \frac{df}{ds}} \right)
$$

$$
= \int\limits_{\underline{s}}^{\overline{s}} ds \, \frac{d\,\delta f}{ds} \cdot \frac{df}{ds} \left(\sqrt{\frac{df}{ds} \cdot \frac{df}{ds}} \right)^{-1} \tag{4.8}
$$

and after integration by parts one obtains

$$
\delta\tau = - \int\limits_{\underline{s}}^{\overline{s}} ds \, \delta f \cdot \frac{d}{ds} \left(\frac{df}{ds} \left(\sqrt{\frac{df}{ds} \cdot \frac{df}{ds}} \right)^{-1} \right). \tag{4.9}
$$

No boundary terms occur on integrating by parts, since $\delta f(\underline{s}) = 0$ and $\delta f(\overline{s}) = 0$.

For a worldline f of extremal length of time, $\delta\tau$ vanishes for all variations δf which vanish at the boundary. So the map f satisfies the differential equation

$$
\frac{d}{ds} \frac{\frac{df}{ds}}{\sqrt{\frac{df}{ds} \cdot \frac{df}{ds}}} = 0, \tag{4.10}
$$

which states that the direction of the tangent is constant. The length of the tangent vector $\frac{df}{ds}$ cannot be fixed by the condition of stationary time, since the time τ, as shown in (4.5), does not depend on the parameterization of the worldline.

That (4.10) is not only sufficient but also necessary for the time to be stationary shows the following consideration. If in the integral (4.9) one component, say the zero component, of the vector which multiplies δf is greater than zero at some value of the parameter, then it is positive in a whole neighborhood, because by assumption it is continuous. If then one chooses the corresponding δf^0 such that it is positive inside this neighborhood and vanishes outside, then $\delta\tau$ is negative and the time τ is not stationary.

If one chooses the parameterization such that the tangent vector has unit length (4.7), then the parameter s coincides up to a constant with the time τ on the worldline and (4.10) states that the tangent vector does not change along the path

$$\frac{\mathrm{d}^2 f}{\mathrm{d}s^2} = 0 \,. \tag{4.11}$$

If moreover, we choose the constant by $f^0(0) = 0$, then the worldline f of extremal time is the straight worldline (2.40) of a free particle

$$f_{\text{free}} : s \mapsto \frac{s}{\sqrt{1 - \mathbf{v}^2}} \begin{pmatrix} 1 \\ \mathbf{v} \end{pmatrix} + \begin{pmatrix} 0 \\ \mathbf{x}(0) \end{pmatrix} . \tag{4.12}$$

The initial position $\mathbf{x}(0)$ and the initial velocity $\mathbf{v}$ are chosen by the initial conditions.

On the straight worldline from A to B the time τ is not only extremal but longer than on each curved worldline, as our discussion of the twin paradox has shown. Time between two events is maximal on the straight worldline which connects them.

4.3 Action Principle

Lift, Jet Functions and Their Change

Each smooth, parameterized curve f (a map of some parameter interval $I \subset \mathbb{R}$ to $\mathbb{R}^d$, where the dimension d is called the number of degrees of freedom) naturally defines its lift of order k

$$\hat{f} : t \mapsto \hat{f}(t) = \left(t, f(t), \frac{\mathrm{d}f}{\mathrm{d}t}(t), \frac{\mathrm{d}^2 f}{\mathrm{d}t^2}(t) \ldots \frac{\mathrm{d}^k f}{\mathrm{d}t^k}(t) \right) \tag{4.13}$$

to a curve in the jet space $\mathscr{J}_k = I \times \mathbb{R}^d \times \mathbb{R}^d \times \mathbb{R}^d \cdots \times \mathbb{R}^d = I \times \mathbb{R}^{d(k+1)}$. Traditionally the coordinates of points of $\mathscr{J}_1$ carry names like (t, x, v) for the time t, the position x and the velocity v. Correspondingly and in a more systematic notation, which indicates the order of differentiation, a point in $\mathscr{J}_k$ is given by coordinates $(t, x, x_{(1)}, x_{(2)} \ldots x_{(k)})$.

Jet space (or a subdomain of it) is the domain of definition of conserved quantities, for example the energy $E(t, \mathbf{x}, \mathbf{v}) = m/\sqrt{1 - \mathbf{v}^2}$ (3.44) of a free particle with mass m is a function of the domain $|\mathbf{v}| < 1$ of $\mathscr{J}_1$. Composed with the lift of a curve f one obtains the energy on the curve in the course of time, $t \mapsto (E \circ \hat{f})(t)$. On the straight worldline of a free particle $E \circ \hat{f}_{\text{free}}$ is constant. Nevertheless, the energy E is not a constant, but a jet function which varies with the velocity. The same applies to each other conserved quantity (3.32). Conserved quantities are jet functions ϕ, which become constant if composed with the lift of a physical path, $\mathrm{d}/\mathrm{d}t\,(\phi \circ \hat{f}_{\text{physical}}) = 0$.

Though constant jet functions are trivial conserved quantities, one should not exclude them by definition from the conserved quantities because one wants them to constitute a vector space.

The derivative d_t of jet functions ϕ with respect to the parameter time t is defined as the usual partial derivative with respect to t and "formally" on the other jet variables, i.e. d_t differentiates each variable $x_{(k)}$ and replaces it by the variable $x_{(k+1)}$. Explicitly, at $(t, x, x_{(1)}, \ldots x_{(k+1)}) \in \mathscr{J}_{k+1}$ the derivative d_t of a function ϕ of $\mathscr{J}_k$ is given by

$$d_t\phi = \left(\partial_t + x_{(1)}^m \partial_{x^m} + x_{(2)}^m \partial_{x_{(1)}^m} \cdots + x_{(k+1)}^m \partial_{x_{(k)}^m} + \ldots\right)\phi\,. \tag{4.14}$$

The index m enumerates the components and runs from 1 to d, the number of the degrees of freedom. We employ Einstein's summation convention. An index pair, such as m in the above equation, designates a sum over the range of its possible values, i.e. the summation sign $\sum_{m=1}^{d}$ is not written but self understood.

The derivative d_t of jet functions is defined such that upon composition with the lift it yields the time derivative of the composed function

$$(d_t\phi) \circ \hat{f} = \frac{d}{dt}\left(\phi \circ \hat{f}\right)\,, \tag{4.15}$$

which, more explicitly, is the chain rule of differentiation,

$$\left(\partial_t + \frac{df^m}{dt}\frac{\partial}{\partial x^m} + \frac{d^2 f^m}{dt^2}\frac{\partial}{\partial x_{(1)}^m} + \cdots + \frac{d^{k+1} f^m}{dt^{k+1}}\frac{\partial}{\partial x_{(k)}^m}\right)\phi\big|_{\left(t, f(t), \frac{df}{dt}(t), \ldots \frac{d^k f}{dt^k}(t)\right)}$$
$$= \frac{d}{dt}\phi\left(t, f(t), \frac{df}{dt}(t), \ldots \frac{d^k f}{dt^k}(t)\right)\,. \tag{4.16}$$

A curve in the space of functions is a map $\lambda \mapsto f_\lambda$ of a parameter λ, which ranges in some interval, to functions f_λ. Jet functions ϕ become functions of the parameter λ when they are evaluated on this curve, i.e. when they are composed with the lift $\hat{f}_\lambda$. According to the chain rule and because partial derivatives can be exchanged, jet functions change on the curve by

$$\frac{\partial}{\partial\lambda}\left((\phi \circ \hat{f}_\lambda)(t)\right) = \frac{\partial}{\partial\lambda}\left(\phi(t, f(t, \lambda), \frac{\partial f}{\partial t}(t, \lambda), \ldots \left(\frac{\partial}{\partial t}\right)^k f(t, \lambda))\right)$$
$$= \left[\left(\frac{\partial}{\partial\lambda}f^m\right)\frac{\partial\phi}{\partial x^m} + \frac{\partial}{\partial t}\left(\frac{\partial}{\partial\lambda}f^m\right)\frac{\partial\phi}{\partial v^m} + \cdots \left(\frac{\partial}{\partial t}\right)^k\right.$$
$$\left.\times\left(\frac{\partial}{\partial\lambda}f^m\right)\frac{\partial\phi}{\partial x_{(k)}^m}\right] \circ \hat{f}_\lambda(t)\,. \tag{4.17}$$

With the notation $\delta x = \frac{\partial}{\partial\lambda}f_\lambda$ we can express the change of ϕ as the differential operator

$$\delta = (\delta x^m)\frac{\partial}{\partial x^m} + (d_t\delta x^m)\frac{\partial}{\partial v^m} + \cdots + ((d_t)^k\delta x^m)\frac{\partial}{\partial x_{(k)}^m} + \cdots \tag{4.18}$$

which acts on the jet function ϕ and which subsequently is evaluated on the curve $\hat{f}$,

$$(\delta\phi) \circ \hat{f}_\lambda = \frac{\partial}{\partial\lambda}(\phi \circ \hat{f}_\lambda) . \tag{4.19}$$

We call $\delta\phi$ the variation or change of ϕ. The operator δ commutes with d_t and therefore is determined by δx, its action on the jet variable x. It depends on the curve f_λ. If there is a curve through each point f in the space of functions such that for all f the corresponding change is a function δx of some jet space $\mathscr{J}_k$, then we call δx and δ local (not to be confused with gauged).

Local Functionals and Euler Derivative

Functionals are maps of functions into the real numbers. For example, the proper time $\tau[f]$ (4.6) shown by an ideal clock in an event B is a functional of the parameterized curve $f : t \mapsto (f^1(t), f^2(t), f^3(t))$ which the clocks traverses in the course of the parameter time until it reaches B.

A local functional S employs its Lagrangian $\mathscr{L}$, a smooth real function of some domain $D \subset \mathscr{J}_k$ of a jet space (we restrict our considerations mainly to $k = 1$)

$$\mathscr{L} : \begin{cases} D \subset \mathscr{J}_1 \to \mathbb{R} \\ (t, x, v) \mapsto \mathscr{L}(t, x, v) \end{cases}, \tag{4.20}$$

to map curves f (with $\hat{f}$ taking values in D) to real numbers $S[f]$ via

$$S[f] = \int dt \, (\mathscr{L} \circ \hat{f})(t) = \int dt \, \mathscr{L}\left(t, f(t), \frac{df}{dt}(t)\right). \tag{4.21}$$

The functional S is called local because it is an integral over the range of parameters t where the integrand at parameter time t depends only on t, the value of f at this time and the value of a finite number if its derivatives, but not on the value of f at some other parameter time t' nor on a Taylor series of f which in $\mathscr{J}_\infty$ represents analytic f at neighboring times $t' \neq t$.

For example, the proper time τ (4.6) is a local functional with Lagrangian $\mathscr{L}(t, \mathbf{x}, \mathbf{v}) = \sqrt{1 - \mathbf{v}^2}$. The ideal clock is insensitive to acceleration in the sense that its Lagrangian is a function of the domain $|\mathbf{v}| < 1$ of the jet space $\mathscr{J}_1$ and does not depend on second or higher derivatives $x_{(k)}$ with $k > 1$.

To investigate a functional S in the neighborhood of some f, we consider it on curves through f, i.e. on one parameter families of curves f_λ with $f_0 = f$. We denote by $\delta f = \partial f_\lambda / \partial\lambda_{|\lambda=0}$ the tangent of the curve. The functional S is differentiable at f, if for all smooth δf, which vanish on the boundary, the derivative $dS[f_\lambda]/d\lambda_{|\lambda=0} =: \delta S[f, \delta f]$ exists and is of the form

$$\delta S[f, \delta f] = \int dt \, \delta f^m(t) \, \frac{\delta S}{\delta f^m(t)} \tag{4.22}$$

where $\frac{\delta S}{\delta f}$, the functional derivative of S at f, is continuous.

The functional derivative $\frac{\delta S}{\delta f}$ is unique: if for continuous h and g and all smooth δf

$$\int dt \, \delta f^m(t) \, h_m(t) = \int dt \, \delta f^m(t) \, g_m(t) \tag{4.23}$$

then $h = g$. Otherwise, if the difference $h_1(t) - g_1(t)$ were positive at some time t it would be positive in a whole neighborhood of this time. Then one could choose a δf^1 which is positive within the neighborhood and vanishes outside and $\delta f^m = 0$ for $m \neq 1$. Then $\int dt \, \delta f^m(t) \, (h_m(t) - g_m(t))$ would be positive and contradict (4.23).

To determine the functional derivative of a local functional (4.21), we differentiate $S[f_\lambda]$ with respect to λ at $\lambda = 0$

$$\frac{d}{d\lambda} S[f_\lambda]_{|\lambda=0} = \int dt \, \frac{\partial}{\partial \lambda}_{|\lambda=0} (\mathscr{L} \circ \hat{f}_\lambda)(t) \overset{4.19}{=} \int dt \, (\delta \mathscr{L} \circ \hat{f})(t) \tag{4.24}$$

where the change of the Lagrangian, assuming it is a function of $\mathscr{J}_1$, is

$$\delta \mathscr{L} = \delta f^m \frac{\partial \mathscr{L}}{\partial x^m} + (d_t \delta f^m) \frac{\partial \mathscr{L}}{\partial v^m}. \tag{4.25}$$

In the last term we shift the derivative away from δf at the cost of a complete time derivative

$$(d_t \delta f^m) \frac{\partial \mathscr{L}}{\partial v^m} = -\delta f^m \left(d_t \frac{\partial \mathscr{L}}{\partial v^m} \right) + d_t \left(\delta f^m \frac{\partial \mathscr{L}}{\partial v^m} \right) \tag{4.26}$$

and obtain

$$\delta \mathscr{L} = \delta f^m \left(\frac{\partial \mathscr{L}}{\partial x^m} - d_t \frac{\partial \mathscr{L}}{\partial v^m} \right) + d_t \left(\delta f^m \frac{\partial \mathscr{L}}{\partial v^m} \right). \tag{4.27}$$

The integral over the complete time derivative contributes only boundary terms to the change of the action

$$\int dt \, d_t \left(\delta f^m \frac{\partial \mathscr{L}}{\partial v^m} \right) \circ \hat{f} \overset{4.15}{=} \int dt \, \frac{d}{dt} \left(\delta f^m \frac{\partial \mathscr{L}}{\partial v^m} \circ \hat{f} \right) = \Big|_{\underline{t}}^{\overline{t}} \delta f^m \frac{\partial \mathscr{L}}{\partial v^m} \circ \hat{f} \tag{4.28}$$

and vanishes if all functions f_λ have the same initial and final values as f, since then the variation δf vanishes at the boundary.

The jet function

$$\frac{\hat{\partial} \mathscr{L}}{\hat{\partial} x^m} := \frac{\partial \mathscr{L}}{\partial x^m} - d_t \frac{\partial \mathscr{L}}{\partial v^m} \tag{4.29}$$

is the Euler derivative of $\mathcal{L}$ in the case that $\mathcal{L}$ is a function of $D \subset \mathcal{J}_1$. The Euler derivative of Lagrangians which depend on higher derivatives $x_{(r)}$ are derived analogously from $\delta\mathcal{L}$ by shifting all derivatives away from $\delta x_{(r)} = (d_t)^r \delta x$,

$$\frac{\hat{\partial}\mathcal{L}}{\hat{\partial}x^m} := \frac{\partial\mathcal{L}}{\partial x^m} + \sum_r (-1)^r (d_t)^r \frac{\partial\mathcal{L}}{\partial x^m_{(r)}} . \qquad (4.30)$$

So the functional derivative at f of a local action (4.21) with Lagrangian $\mathcal{L}$ with respect to variations which vanish on the boundary is the Euler derivative of the Lagrangian, composed with the lift of f,

$$\frac{\delta S}{\delta f^m} = \frac{\hat{\partial}\mathcal{L}}{\hat{\partial}x^m} \circ \hat{f} . \qquad (4.31)$$

The Principle of Stationary Action

The equations, which determine the time evolution of elementary physical systems,[1] state that a corresponding local functional, the action of the system, is stationary at the physical paths. From its initial position the system traverses in the course of time that curve to the final position which makes the action stationary in comparison to all other conceivable curves from the same initial position to the same final position.

For example, a free particle with mass m traverses straight worldlines and therefore worldlines which extremize the action (4.21) with Lagrangian[2]

$$\mathcal{L}_{\text{free}}(t, \mathbf{x}, \mathbf{v}) = -m\sqrt{1 - \mathbf{v}^2} . \qquad (4.32)$$

The action S is stationary at f, if its variational derivative vanishes at f. So the Euler derivative (4.29) of the Lagrangian vanishes on the physical path f_{physical}

$$\frac{\hat{\partial}\mathcal{L}}{\hat{\partial}x^m} \circ \hat{f}_{\text{physical}} = 0 . \qquad (4.33)$$

These are the Euler-Lagrange equations which single out the physical paths f_{physical} from all other conceivable paths from the same initial to the same final position.

The Euler-Lagrange equations of the free relativistic particle state, that its momentum (3.42) is constant,

---

[1] These are systems with no friction and with only such constraints on the jet variables $x_{(r)}$, which follow as time derivatives of constraints on the positions.

[2] We choose the sign and the normalization of the Lagrangian of the free relativistic particle such that to first order in v^2 it agrees with the Newtonian Lagrangian $\mathcal{L}_{\text{Newton}} = E_{\text{kin}} = \frac{1}{2}m\mathbf{v}^2$ up to an irrelevant constant. Then the conserved quantities of the relativistic and the nonrelativistic particles coincide in leading order.

$$-\frac{d}{dt}\frac{m\,\mathbf{v}}{\sqrt{1-\mathbf{v}^2}} = 0\,, \quad \mathbf{v} = \frac{d\mathbf{f}_{physical}}{dt}\,. \tag{4.34}$$

Consequently its worldline is straight.

In Newtonian mechanics, the Lagrangian is the difference

$$\mathscr{L}(t, x, v) = E_{kin} - E_{pot} \tag{4.35}$$

of kinetic and potential energy. In particular, the Lagrangian of the harmonic oscillator, a particle with mass m on a spring with constant κ, is

$$\mathscr{L}_{oscillator}(t, x, v) = \frac{1}{2}m\,v^2 - \frac{1}{2}\kappa x^2\,. \tag{4.36}$$

Its Euler derivative is $-m\,x_{(2)} - \kappa\,x$. The Euler-Lagrange equation $-m\,\ddot{f} - \kappa f = 0$ has the solution $f : t \mapsto a\cos(\omega t + \varphi)$, where $\omega = \sqrt{\kappa/m}$ is 2π times the frequency. The amplitude a and the phase φ are chosen by the initial conditions.

The Euler-Lagrange equations (4.33) hold in all coordinate systems, for the principle of stationary action does not make use of the choice of coordinates which describe the path. If we view the coordinates x as functions $x(t, y)$ of other coordinates y and if we convert the Lagrangian

$$\tilde{\mathscr{L}}(t, y, w) = \mathscr{L}\left(t, x(t, y), \frac{\partial x}{\partial t} + \frac{\partial x}{\partial y}w\right), \tag{4.37}$$

the Euler-Lagrange equations hold in y-coordinates if and only if they are satisfied in x-coordinates

$$\frac{\hat{\partial}\tilde{\mathscr{L}}}{\hat{\partial}y^m} = \frac{\partial x^n}{\partial y^m}\frac{\hat{\partial}\mathscr{L}}{\hat{\partial}x^n}\,. \tag{4.38}$$

This follows from (4.22), $\delta x = \frac{\partial x}{\partial y}\delta y$ and (4.31).

A function $\mathscr{L}$ of the jet variables is a time derivative if and only if its Euler derivative vanishes as a function of the jet variables. It is simple to verify that the Euler derivative vanishes if $\mathscr{L} = d_t K = \partial_t K + v^m\frac{\partial}{\partial x^m}K$ is a time derivative. To show the converse, we write the Lagrangian as an integral over its derivative

$$\mathscr{L}(t, x, v) = \mathscr{L}(t, 0, 0) + \int_0^1 d\lambda\,\frac{\partial}{\partial\lambda}\mathscr{L}(t, \lambda x, \lambda v)\,. \tag{4.39}$$

The derivative of the Lagrangian with respect to λ is (4.27)

$$\frac{\partial\mathscr{L}}{\partial\lambda}(t, \lambda x, \lambda v) = x^m \left.\frac{\hat{\partial}\mathscr{L}}{\hat{\partial}x^m}\right|_{(t,\lambda x,\lambda v)} + d_t\left(x^m \left.\frac{\partial\mathscr{L}}{\partial v^m}\right|_{(t,\lambda x,\lambda v)}\right)\,. \tag{4.40}$$

Therefore the Lagrangian can be written as

$$\mathscr{L}(t, x, v) = x^m \int\limits_0^1 d\lambda \left.\frac{\hat{\partial}\mathscr{L}}{\hat{\partial}x^m}\right|_{(t,\lambda x,\lambda v)} + d_t \left(x^m \int\limits_0^1 d\lambda \frac{\partial\mathscr{L}}{\partial v^m} + \int\limits^t dt' \mathscr{L}(t', 0, 0) \right).$$

(4.41)

This is a total time derivative if the Euler derivative of the Lagrangian vanishes identically in the jet variables for all $(t, \lambda x, \lambda v)$.

4.4 Symmetries and Conserved Quantities

Transformation of Functions

A group is a set G with an associative product

$$\begin{array}{c} G \times G \to G \\ a \quad b \quad \mapsto ab \end{array}, \quad a(bc) = (ab)c,$$

(4.42)

a unit element e, which leaves all group elements a unchanged,

$$ea = ae = a,$$

(4.43)

where each element has an inverse,

$$a^{-1}a = aa^{-1} = e.$$

(4.44)

For instance, the nth roots of 1 constitute a group, the cyclic group with n elements,

$$\mathbb{Z}_n = \{1, z, z^2 \ldots z^{n-1}\}, \quad z = e^{\frac{2\pi i}{n}}.$$

(4.45)

Also the invertible maps of each set $\mathscr{M}$ to itself, the transformations $T : \mathscr{M} \to \mathscr{M}$, constitute a group with successive application as group product.

If a subset H of a group G contains each product and each inverse of its elements, then H is a group itself, a subgroup of G. For example, the linear transformations L of a d-dimensional complex vector space V form the general linear group $\mathrm{GL}(d, \mathbb{C})$. The determinants of volume preserving transformations L have the special value $\det L = 1$. Volume preserving linear transformations are called unimodular and constitute the subgroup $\mathrm{SL}(d, \mathbb{C})$ of special linear transformations. If the vector spaces are real, the corresponding transformation groups are called $\mathrm{GL}(d, \mathbb{R})$ and $\mathrm{SL}(d, \mathbb{R})$.

Some of the transformations of $\mathscr{M}$ may leave some subset or some property invariant. Then they are called symmetries of the subset or symmetries of the property, e.g. the linear transformations of a vector space are the symmetries of the set of straight lines. The set of symmetries (of some considered mathematical structure) is a subgroup of the group of transformations, e.g. rotations form the subgroup of linear transformations of a Euclidean space, which leave length invariant.

The same transformation may act on different objects, e.g. a rotation may rotate a point, a rigid body or a force. Then one has to distinguish between the transformations of the different objects and one has also to identify them in some sense. This is achieved by the concept of a realization of a group. A realisation of a group G is a map N which maps the group elements $g \in G$ to the transformations N_g of some manifold $\mathcal{N}$, such that the unit element corresponds to the identity and successive transformations yield the transformation corresponding to the group product,

$$N_{g2} \circ N_{g1} = N_{g2\,g1} \,. \tag{4.46}$$

If M_g is another realization of G, then N_g and M_g are in a loose sense the same transformations but strictly speaking they are different realizations of the same group.

The realization of a group by linear transformations is called a representation.

If a group is realized on $\mathcal{M}$ by M_g and on $\mathcal{N}$ by N_g, then it is also realized on the cartesian product $\mathcal{M} \times \mathcal{N}$, the set of pairs (x, y), $x \in \mathcal{M}$, $y \in \mathcal{N}$, by $M_g \times N_g : (x, y) \mapsto (M_g x, N_g y)$.

Each map $f : \mathcal{M} \to \mathcal{N}$ is a subset[3] of the cartesian product, $f \subset \mathcal{M} \times \mathcal{N}$, with the property that for each $x \in \mathcal{M}$ it contains precisely one pair (x, y). This pair defines $f(x)$ by $y = f(x)$. In other words, f is a section of the product which cuts each fiber $\{x\} \times \mathcal{N}$ once.

The same function is also the set of pairs $(M_g^{-1} x, f(M_g^{-1} x))$, as M_g is invertible. These pairs are mapped by $M_g \times N_g$ to the pairs $(x, N_g\, f(M_g^{-1} x)$. So the realization of a group G as transformation group on a manifold $\mathcal{M}$ and a manifold $\mathcal{N}$ maps the space $\mathcal{F}$ of functions f from $\mathcal{M}$ to $\mathcal{N}$ to itself by the transformation, which is adjoint to N_g and M_g

$$\mathrm{ad}_g(f) = N_g \circ f \circ M_g^{-1} \,. \tag{4.47}$$

If N_g is the identity for all g, then the functions f are called scalar fields.

The adjoint transformations ad_g are a realization of the group G on the space $\mathcal{F}$ of functions from $\mathcal{M}$ to $\mathcal{N}$. They consist of a left multiplication by a realization N_g, which satisfies (4.46), together with a right multiplication with the inverse of a realization M_g. This is also a realization, $M_{g1}^{-1} \circ M_{g2}^{-1} = (M_{g2} \circ M_{g1})^{-1} = M_{g2\,g1}^{-1}$.

There are also other realizations of G on function spaces $\mathcal{F}$. If G acts on $\mathcal{M}$ and a subgroup $H \subset G$ leaves a point $\underline{x} \in \mathcal{M}$ invariant, then $\mathcal{M}$ consists at least of the points $x \in G/H$ which are obtained by some transformation $\pi(x) \in G$ applied to $\underline{x}$. (Locally there exist bijective maps $\pi : U \subset G/H \to \pi(U) \subset G$.) If H is realized by transformations N_h of $\mathcal{N}$, then this induces ind_g, the induced realization of G on the maps f from $\mathcal{M} = G/H$ to $\mathcal{N}$, by $(\mathrm{ind}_g f)(gx) = N_{h(g,x)} \circ f(x)$ where

---

[3] To call this set the graph of the function f spoils the opportunity to define what the function f is.

$h(g, x) \in H$ is defined by $h(g, x) = \pi(gx)^{-1} g\pi(x)$ and satisfies $h(g_2 g_1, x) = h(g_2, g_1 x) h(g_1, x)$.

Infinitesimal Symmetries

In mechanics we deal with curves $f : t \mapsto x = f(t)$ and continuous transformations such as translations by a multiple of a vector c

$$T_\alpha f = f + \alpha c \tag{4.48}$$

or a temporal translation by (minus) some time α

$$(T_\alpha f)(t) = f(t + \alpha) \tag{4.49}$$

or a rotation by an angle α

$$T_\alpha \begin{pmatrix} f^1 \\ f^2 \end{pmatrix} = \begin{pmatrix} \cos\alpha & -\sin\alpha \\ \sin\alpha & \cos\alpha \end{pmatrix} \begin{pmatrix} f^1 \\ f^2 \end{pmatrix}. \tag{4.50}$$

In these examples of continuous transformations the transformation parameter α parameterizes a curve in a group. It is chosen such that for $\alpha = 0$ the curve passes the unit element, $T_0 f = f$. The tangent vector to the curve $T_\alpha f$ at the unit element is called δf, the infinitesimal transformation of f,

$$\delta f = \partial_\alpha \big|_{\alpha=0} T_\alpha f . \tag{4.51}$$

We have $\delta f = c$ for spatial translations, $\delta f = \frac{df}{dt}$ for temporal translations and $\delta(f^1, f^2) = (-f^2, f^1)$ for rotations. In these examples the change of all curves f under an infinitesimal transformation is local (compare page 62), i.e. a jet function δx evaluated on the lift of the curve, $\delta f = \delta x \circ \hat{f}$.

The corresponding change of the Lagrangian of a local action is (assuming it is a function of $\mathscr{J}_1$ because otherwise the energy cannot be bounded from below)

$$\delta\mathscr{L} = \delta x^m \frac{\partial\mathscr{L}}{\partial x^m} + (d_t \delta x^m) \frac{\partial\mathscr{L}}{\partial v^m} \overset{4.27}{=} \delta x^m \frac{\hat{\partial}\mathscr{L}}{\hat{\partial} x^m} + d_t \left(\delta x^m \frac{\partial\mathscr{L}}{\partial v^m} \right). \tag{4.52}$$

If the corresponding action (4.21) changes by boundary terms only, i.e. if $\delta\mathscr{L}$ is the derivative of some jet function K,

$$\delta\mathscr{L} + d_t K = 0 , \tag{4.53}$$

then we call δ or δx an infinitesimal symmetry of the action.

From the definition of an infinitesimal symmetry of the action (4.53), and from (4.52) one concludes

$$\delta x^m \frac{\hat{\partial}\mathscr{L}}{\hat{\partial}x^m} + d_t \mathscr{Q} = 0 \,, \tag{4.54}$$

$$\mathscr{Q} = K + \delta x^m \frac{\partial \mathscr{L}}{\partial v^m} \,. \tag{4.55}$$

Equation (4.54) subsumes the

Noether Theorem: *To each infinitesimal symmetry of the action there corresponds a conserved quantity. Vice versa, to each conserved quantity there corresponds an infinitesimal symmetry of the action* [24].

The theorem holds, because the physical paths satisfy the equations of motion $\frac{\hat{\partial}\mathscr{L}}{\hat{\partial}x^m} \circ \hat{f}_{\text{physical}} = 0$ (4.33). Consequently the Noether charge $\mathscr{Q}$ is a conserved quantity, $d/dt\,(\mathscr{Q} \circ \hat{f}_{\text{physical}}) = 0$.

Conversely, to each conserved quantity there corresponds an infinitesimal symmetry of the action. This is true because by definition a jet function $\overline{\mathscr{Q}}$ is a conserved quantity if the time derivative of $\overline{\mathscr{Q}} \circ \hat{f}_{\text{physical}}$ vanishes due to the equations of motion, i.e. if $d_t \overline{\mathscr{Q}}$ can be written as a multiple of the Euler derivative of the Lagrangian and of the derivatives of the Euler derivative with some jet functions r_0 and r_1,

$$d_t \overline{\mathscr{Q}} + r_0^m \frac{\hat{\partial}\mathscr{L}}{\hat{\partial}x^m} + r_1^m d_t \frac{\hat{\partial}\mathscr{L}}{\hat{\partial}x^m} = 0 \,. \tag{4.56}$$

If we combine the terms with the product rule and redefine the conserved quantity by terms $r_1^m \frac{\hat{\partial}\mathscr{L}}{\hat{\partial}x^m}$ which vanish on physical paths then it is related to an infinitesimal symmetry $\delta x = r_0 - d_t r_1$ in standard form (4.54),

$$d_t \left(\overline{\mathscr{Q}} + r_1^m \frac{\hat{\partial}\mathscr{L}}{\hat{\partial}x^m} \right) + \left(r_0^m - d_t r_1^m \right) \frac{\hat{\partial}\mathscr{L}}{\hat{\partial}x^m} = 0 \,. \tag{4.57}$$

One cannot require infinitesimal symmetries δx to vanish on the boundary. This restriction would exclude translations and rotations. We also do not require that infinitesimal symmetries can be integrated to finite transformations. Such requirements would spoil the correspondence of conserved quantities (5.150).

The Noether theorem is important because often symmetries of the action are evident and can be seen as a geometric property of the Lagrangian, e.g. that it is invariant under translations or rotations or independent of the parameter time t.

Conserved quantities are crucial for the integrability of the equations of motion, i.e. whether the solutions can be obtained by integration of given functions or by solving for implicitly given functions. If the equations of motion apply to d degrees of freedom, the equations are integrable if and only if there are d independent conserved quantities $\mathcal{Q}_1 \ldots \mathcal{Q}_d$, whose corresponding infinitesimal transformations $\delta_1, \ldots \delta_d$, when applied successively, can be interchanged just like translations $\delta_i \delta_j = \delta_j \delta_i$. If one modifies integrable equations of motion by additional terms, such perturbations, even if they are small, lead to chaotic paths which, though they have small measure, are dense in the space of all paths. The derivation and discussion of these important results fill books [1, 6, 21], to which we refer for a thorough presentation.

Symmetries of the action are not only important because they are related to conserved quantities but also because the transformed solutions to the equations of motion are solutions which in some cases may turn out to be unknown previously. In general relativity one obtains the gravitational field of a uniformly moving mass from the gravitational field of a mass at rest by a Lorentz transformation. Thereby one proves that rapid motion does not turn a mass into a black hole.

Transformations which map solutions of the equations of motion to solutions are not necessarily symmetries of the action, e.g. $f : t \mapsto -\frac{1}{2} g t^2 + v_0 t + x_0$, the solutions of the equations of motion $-\ddot{f} - g = 0$ of a vertically falling particle, are transformed into each other by $T_\alpha f : t \to e^{2\alpha} f(e^{-\alpha} t)$. Nevertheless the infinitesimal transformation $\delta x = 2x - t\,v$ does not leave the Lagrangian $\mathcal{L} = \frac{1}{2} m v^2 - mgx$ invariant up to a derivative, as the Euler derivative of $\delta \mathcal{L} = d_t(-t\,\mathcal{L}) + 5\mathcal{L} + 2 mgx$ does not vanish.

Energy and Momentum

The Lagrangian $\mathcal{L}(t, \mathbf{x}, \mathbf{v}) = -m\sqrt{1 - \mathbf{v}^2}$ (4.32) of a free relativistic particle is invariant under Poincaré transformations, i.e. under spatial and temporal translations, rotations and boosts.

The conserved quantity (4.55) which corresponds to the infinitesimal translation $\delta \mathbf{x} = \mathbf{c}$ by a constant vector $\mathbf{c}$,

$$c^i \frac{\partial \mathcal{L}}{\partial v^i} = \mathbf{c} \cdot \mathbf{p}, \quad \mathbf{p} = \frac{m\,\mathbf{v}}{\sqrt{1 - \mathbf{v}^2}}, \tag{4.58}$$

is by definition (and in agreement with (3.42)) the momentum $\mathbf{p}$ in direction of the vector $\mathbf{c}$. Translation invariance of the action corresponds to the conservation of momentum.

If the Lagrangian depends only on the velocity, not on the coordinate of a degree of freedom, then the variable is called cyclic. Then the Lagrangian is invariant under translation in this direction, e.g. under $(x^1, x^2, \ldots x^d) \mapsto (x^1 + \alpha, x^2, \ldots x^d)$ and under the infinitesimal transformation $\delta x = (1, 0, \ldots 0)$. The Noether charge (4.55) is the canonically conjugate momentum $\frac{\partial \mathcal{L}}{\partial v^1}$ of the cyclic variable. The Euler Lagrange equations just state that it is conserved,

$$\frac{\partial \mathcal{L}}{\partial x^1} = 0 \wedge \left(\frac{\partial \mathcal{L}}{\partial x^1} - d_t \frac{\partial \mathcal{L}}{\partial v^1} \right) \circ f_{\text{physical}} = 0 \Rightarrow \frac{d}{dt} \left(\frac{\partial \mathcal{L}}{\partial v^1} \circ f_{\text{physical}} \right) = 0. \tag{4.59}$$

Temporal translation (4.49) yields the infinitesimal transformation

$$\delta x = v. \tag{4.60}$$

This is an infinitesimal symmetry of the action whenever the Lagrangian $\mathscr{L}(t, x, v)$ does not depend on t, since $\partial_t \mathscr{L} = 0$ implies

$$\delta \mathscr{L} = v^m \, \partial_{x^m} \mathscr{L} + (d_t v^m) \, \partial_{v^m} \mathscr{L} \overset{4.14}{=} d_t \mathscr{L} - \partial_t \mathscr{L} = d_t \mathscr{L} , \tag{4.61}$$

i.e. (4.53) with $K = -\mathscr{L}$. The Noether charge (4.55) corresponding to the invariance under time translation is by definition the energy E,

$$E = v^m \frac{\partial}{\partial v^m} \mathscr{L} - \mathscr{L} . \tag{4.62}$$

The energy is conserved if the Lagrangian does not depend explicitly on the time.

In agreement with (3.44) the energy of a free particle with Lagrangian (4.32) is

$$E(t, \mathbf{x}, \mathbf{v}) = \frac{m}{\sqrt{1 - \mathbf{v}^2}} . \tag{4.63}$$

The definition (4.62) of the energy contains the recipe (4.35) for the New-tonian Lagrangian. The operator $v \partial_v$ counts the degree of homogeneity in veloc-ities, $v \partial_v (v)^n = n (v)^n$, where the superscript denotes an exponent for once. If one decomposes the Lagrangian and the energy into pieces L_n and E_n, which are homo-geneous of degree n in the velocities, then by (4.62) $E = \sum E_n = \sum_n (n - 1) L_n$ and $\mathscr{L} = \sum_{n \neq 1} E_n / (n - 1) + L_1$. A part of the Lagrangian, which is linear in the velocities $L_1(x, v) = q v^i A_i(x)$ does not contribute to the energy but adds a magnetic force $q v^j (\partial_{x^i} A_j - \partial_{x^j} A_i)$ to the Euler derivative and to the equations of motion. In Newtonian mechanics the energy is the sum of kinetic energy E_{kin}, which is quadratic in the velocity, $n = 2$, and potential energy E_{pot}, which is independent of the velocity, $n = 0$, so the Lagrangian is $\mathscr{L} = E_{\text{kin}} - E_{\text{pot}} + L_1$.

In Hamiltonian mechanics the energy (4.62) defines the Hamiltonian

$$\mathscr{H}(x, p) = v^m \, p_m - \mathscr{L} , \quad p_m = \partial_{v^m} \mathscr{L} , \tag{4.64}$$

as a function of the phasespace variables, the position x and the canonically conjugate momentum $p = \partial_v \mathscr{L}$, rather than the jet variables x and v.

The motion of particles in phasespace is an area preserving map with respect to the measure $d p_m \, d x^m - d H \, d t$ [1]. This important geometric property is basic for the Kolmogorov-Arnold-Moser theorem [21] and the conclusion that even small perturbations of integrable motions lead to chaotic curves [6] which are dense in the space of all curves though they can have small measure.

If in a one-dimensional motion the energy, a function of $\mathscr{I}_1$, is conserved, i.e. if $\frac{d}{dt}E \circ f_{\text{physical}} = 0$, then the parameter time $t(x)$ at which the particle passes the point x can be calculated as an integral and the path $x = f(t)$ can be obtained as the inverse function of $t(x)$. For example, if the energy is of the form

$$E(x, v) = \frac{1}{2}m v^2 + V(x) \tag{4.65}$$

with some potential V, then we solve for the velocity v and, choosing the sign of v, obtain

$$v(x, E) = \sqrt{\frac{2}{m}(E - V(x))} \tag{4.66}$$

while x increases. By the implicit function theorem, an analogous solution $v(x, E)$ exists for general $E(x, v)$ in the neighborhood of each possible value of the energy where the energy varies with the velocity. On the physical path $v \circ \hat{f} = df/dt$ and $E \circ \hat{f}$ is a constant number. The derivative of the inverse function $t(x)$ at x is the reciprocal $\frac{dt}{dx}|_{x=f(t)} = (\frac{df}{dt})^{-1}|_t$

$$\frac{dt}{dx}(x) = \left(\sqrt{\frac{2}{m}(E - V(x))}\right)^{-1}, \tag{4.67}$$

so the wanted function $t(x)$ is an integral over a known integrand

$$t(x) - t(\underline{x}) = \int_{\underline{x}}^{x} dx' \frac{1}{\sqrt{\frac{2}{m}(E - V(x'))}}. \tag{4.68}$$

All solvable equations of motion are solvable because they possess sufficiently many conserved quantities and because the conserved quantities determine the solutions as inverse functions of integrals over known functions just as in the example of one-dimensional motion.

Angular Momentum and Energy Weighted Position

If the action is invariant under rotations around an axis (4.50), the corresponding Noether charge is by definition the angular momentum in the direction of this axis.

A rotation D is a linear transformation $\mathbf{x} \mapsto D\mathbf{x}$, where D is orthogonal (6.6), $D^T D = 1$. Differentiating this relation for a one parameter set of rotations D_λ with $D_0 = 1$ at $\lambda = 0$ one concludes that each infinitesimal rotation $r = \partial_\lambda D_\lambda|_{\lambda=0}$ is an antisymmetric matrix, $r^T = -r$. In three dimensions the linear map r maps each vector $\mathbf{x}$ to the vector product with a vector $\mathbf{r} = (r^1, r^2, r^3)$,

$$r = \begin{pmatrix} & -r^3 & r^2 \\ r^3 & & -r^1 \\ -r^2 & r^1 & \end{pmatrix}, \quad r\mathbf{x} = \mathbf{r} \times \mathbf{x}. \tag{4.69}$$

The Langrangian (4.32) is invariant under each infinitesimal rotation $\delta\mathbf{x} = \mathbf{r} \times \mathbf{x}$ and the corresponding change of the velocities $\delta\mathbf{v} = d_t\delta\mathbf{x} = \mathbf{r} \times \mathbf{v}$. So the corresponding Noether charge (4.55) is

$$\mathbf{r} \cdot \mathbf{L} = (\mathbf{r} \times \mathbf{x}) \cdot \frac{m\,\mathbf{v}}{\sqrt{1-v^2}} = \mathbf{r} \cdot (\mathbf{x} \times \mathbf{p}). \tag{4.70}$$

Here we have used $(\mathbf{a} \times \mathbf{b}) \cdot \mathbf{c} = \mathbf{a} \cdot (\mathbf{b} \times \mathbf{c})$. The Noether charge defines the angular momentum $\mathbf{L}$ of a relativistic particle. As in Newtonian mechanics it is the vector product of the position $\mathbf{x}$ with the momentum $\mathbf{p}$

$$\mathbf{L} = \mathbf{x} \times \mathbf{p}. \tag{4.71}$$

Each Lorentz transformation Λ of spacetime can be decomposed into a rotation and a symmetric matrix $L_P = (L_P)^T$ (6.35), the Lorentz boost. The conserved quantities which are related to rotations have already been considered. It remains to investigate boosts and their Noether charges. To lowest order in the transformation parameter v, the boost (3.7) in x-direction changes t by $-vx$ and x by $-vt$ and leaves y and z invariant. More generally a boost in an arbitrary direction changes t and $\mathbf{x}$ by an infinitesimal transformation $\delta_t = \mathbf{c} \cdot \mathbf{x}$ and $\delta_{\mathbf{x}} = \mathbf{c}\,t$.

The corresponding change of curves $\mathbf{f} : \mathbb{R} \to \mathbb{R}^3$, which particles follow in the course of time, is more complicated than (4.47) because Lorentz boosts do not act on the product space $\mathbb{R} \times \mathbb{R}^3$ by a product of representations. They map the set of pairs $(t, \mathbf{x}) = (t, \mathbf{f}(t))$ (which define the map $\mathbf{f}$) to lowest order to the set $(t + \delta_t, \mathbf{x} + \delta_{\mathbf{x}}) = (t + \delta_t, \mathbf{f}(t) + \delta_{\mathbf{x}})$ which in this order is the same as the set $(t, \mathbf{f}(t) - \frac{d\mathbf{f}}{dt}\delta_t + \delta_{\mathbf{x}})$, i.e. the function $\mathbf{f}$ changes under a boost by the jet function

$$\delta\mathbf{x}(t, \mathbf{x}, \mathbf{v}) = -\mathbf{v}\delta_t + \delta_{\mathbf{x}} = -\mathbf{v}(\mathbf{c} \cdot \mathbf{x}) + \mathbf{c}\,t \tag{4.72}$$

evaluated on the lift, $\delta\mathbf{f} = \delta\mathbf{x} \circ \hat{\mathbf{f}}$. Correspondingly the velocities change by

$$\delta\mathbf{v}(t, \mathbf{x}, \mathbf{v}, \mathbf{x}_{(2)}) = d_t\delta\mathbf{x} = \mathbf{c} - \mathbf{x}_{(2)}(\mathbf{c} \cdot \mathbf{x}) - \mathbf{v}(\mathbf{c} \cdot \mathbf{v}). \tag{4.73}$$

This is an infinitesimal symmetry of the Lagrangian $\mathscr{L} = -m\sqrt{1-\mathbf{v}^2}$ (4.32),

$$-\delta\sqrt{1-\mathbf{v}^2} = \frac{\mathbf{v} \cdot \delta\mathbf{v}}{\sqrt{1-v^2}} = \left(\mathbf{c} \cdot \mathbf{v}\sqrt{1-\mathbf{v}^2} - \mathbf{c} \cdot \mathbf{x}\frac{\mathbf{x}_{(2)} \cdot \mathbf{v}}{\sqrt{1-\mathbf{v}^2}}\right)$$
$$= -d_t\left(\mathbf{c} \cdot \mathbf{x}\sqrt{1-\mathbf{v}^2}\right). \tag{4.74}$$

The Noether charge (4.55) is the product of $\mathbf{c}$ with the energy weighted initial position $(\mathbf{K} \circ \hat{f}_{\text{free}} = E \, \mathbf{f}_{\text{free}}(0))$,

$$\mathbf{K} = m \, \frac{\mathbf{x} - \mathbf{v} t}{\sqrt{1 - \mathbf{v}^2}} = \mathbf{x} E - t \mathbf{p} \, . \tag{4.75}$$

Its components and the components of the angular momentum $\mathbf{L}$ are the antisymmetrized products of the components of the four-vectors $x = (t, \mathbf{x})$ and $p = (E, \mathbf{p})$

$$M^{mn} = x^m p^n - x^n p^m \, , \quad M^{mn} = -M^{nm} \, , \quad m, n \in \{0, 1, 2, 3\} \, . \tag{4.76}$$

Under Poincaré transformations $x' = \Lambda x + a$ the four-momentum p transforms as a difference vector $p' = \Lambda p$ (3.39), so M and p transform linearly and reducibly as

$$M'^{mn} = \Lambda^m{}_k \Lambda^n{}_l M^{kl} + (a^m \Lambda^n{}_l - a^n \Lambda^m{}_l) p^l \, , \quad p'^k = \Lambda^k{}_l p^l \, . \tag{4.77}$$

In particular, the transformation of the angular momentum under translations is the Huygens-Steiner theorem $\mathbf{L}' = \mathbf{L} + \mathbf{a} \times \mathbf{p}$.

The Lagrangian for N free particles is simply the sum $-\sum_i m_i \sqrt{1 - \mathbf{v}_{[i]}^2}$ of the Lagrangians (4.32) of the individual particles which traverse paths $\mathbf{f}_{[i]} : t \mapsto \mathbf{f}_{[i]}(t)$. Trivially, the total four-momentum (or momentum for short)

$$p = \sum_{i=1}^{N} p_{[i]} \tag{4.78}$$

remains conserved, as each individual momentum is a conserved quantity. However, by the Noether theorem the total momentum remains a conserved quantity of interacting particles if the interaction is invariant under translations of space and time. If one can neglect the contribution of the interaction to the momentum when the particles are far apart, then this conserved momentum is given by the sum (4.78) of the momenta of the individual, sufficiently distant particles.

That the energy and spatial momentum of the real particles are conserved has been verified in all relevant experiments and observations. By the Noether theorem this means that the action is invariant under translations in spacetime.

However, it required a "desperate hypothesis" by Pauli who introduced a neutrino as an additional bookkeeping entry to account for the energy-momentum balance in the beta decay of nuclei. Decades later the neutrinos, more precisely three kinds of neutrinos by now, were detected and their postulated properties were verified experimentally.

Energy conservation does not hold in general relativity, if the metric varies in the course of time. For instance, in the expanding universe the background radiation cools down while the draining energy is not transformed into other measurable energy. The loss of energy corresponds to a lack of invariance under time translations: the big bang distinguishes a particular time some 13×10^9 years ago.

The invariance of the action under boosts implies that the sum of $\mathbf{K} = \mathbf{x}E - t\mathbf{p}$ over the individual particles is time-independent on the physical path, i.e. composed with $\hat{f}_{\text{physical}}$. Therefore the center of energy $\mathbf{X} = \sum \mathbf{x}E / \sum E$ moves linearly and uniformly with the velocity $\mathbf{V} = \sum \mathbf{p}/\sum E$

$$\frac{\sum(\mathbf{x}E - t\mathbf{p})}{\sum E} = \frac{\sum \mathbf{x}(0)\, E}{\sum E} \Leftrightarrow \mathbf{X}(t) = \mathbf{V}\, t + \mathbf{X}(0) \,. \qquad (4.79)$$

4.5 Interlude in Linear Algebra

Each linear map M of a (real) vector space V to a vector space W is determined by its action on a basis $e_1, e_2 \ldots e_d$, because each vector $v \in V$ is a linear combination $v = e_m v^m$ with real components $v^1, v^2 \ldots v^d$ and by linearity $M(e_m v^m) = M(e_m)\, v^m$. The image $M(e_m) = e'_n M^n{}_m$, where $e'_1, e'_2 \ldots$ are the basis vectors of W, is a vector with components $M^n{}_m$, the element of the matrix M (in the basis e and e') in the row n (first index) and the column m (second index). The mth column of a matrix contains the components of the image of the basis vector e_m. The image Mv has components $(Mv)^n = M^n{}_m v^m$. The set of linear maps M from V to W is a vector space, because one can naturally add and scale linear maps.

In particular each linear map l of V to $\mathbb{R}$ is given by its components $l_n = l(e_n)$, $l(v) = l(e_n v^n) = l(e_n) v^n = l_n v^n$. The bilinear map $(l, v) \mapsto l(v) = l_1 v^1 + l_2 v^2 + \cdots$ is sometimes called the scalar product of l with v. However, l is a vector not of V but of the dual vector space V^* which makes a difference because vectors $v \in V$ and their linear maps $l \in V^*$ transform differently, e.g. under Lorentz boosts with the opposite sign of the velocity.

Each l is a linear combination $l_m f^m$ of the dual basis $f^1, f^2 \ldots f^d$, which map vectors v to their components, $f^m(v) = v^m$, in particular $f^m(e_n) = \delta^m{}_n$, where $\delta^m{}_n$ are the components of the Kronecker δ,

$$\delta^m{}_n = \begin{cases} 1 & \text{if } m = n \\ 0 & \text{if } m \neq n \end{cases} . \qquad (4.80)$$

Each linear map M of a vector space V to a vector space W defines the transposed map M^T from W^* to V^* by composition, also called pullback or right multiplication,

$$l \mapsto M^T l = l \circ M \,. \qquad (4.81)$$

The matrix M^T is M reflected along the diagonal: the rows of M^T contain the columns of M, $M_n^{T\,m} = M^m{}_n$, $(M^T l)_m = l(M e_m) = l(e_n M^n{}_m) = l_n M^n{}_m = M^T{}_m{}^n l_n$.

If linear transformations D act on the vector space V, then, assuming D does not transform the target space $\mathbb{R}$, the linear maps l from V to $\mathbb{R}$ transform by right multiplication with D^{-1} (4.47)

$$l \mapsto l \circ D^{-1} = D^{T-1}l \,.\tag{4.82}$$

This transformation is contragredient to the transformation of vectors in the sense, that l', the transformed l, applied to v', the transformed vector v, gives the same result as before the transformation, $l'(v') = (l \circ D^{-1})(Dv) = l(D^{-1}Dv) = l(v)$.

The contragredient transformation D^{T-1} is a representation of a group G if D is: $D_{g_2} D_{g_1} = D_{g_2 g_1}$ implies $D_{g_2}^{T-1} D_{g_1}^{T-1} = (D_{g_1}^{-1} D_{g_2}^{-1})^T = D_{g_2 g_1}^{T-1}$.

We indicate the transformation property of vectors with the index position of their components. Vector components have upper indices and transform with D, components of dual vectors have lower indices and transform with the contragredient representation D^{T-1},

$$v'^m = D^m{}_n v^n \,, \quad l'_m = D^{T-1}{}_m{}^n l_n \,.\tag{4.83}$$

The sum of a lower with an upper index is invariant, $l'(v') = l'_m v'^m = l_m v^m = l(v)$.

Tensor is the collective name for maps T which are linear in several vector arguments. A tensor which depends on no vector argument is called scalar, a tensor l with one vector argument is called a dual vector and also a vector v is a tensor, because it defines a linear map of dual vectors l to the numbers $l(v)$. Examples of tensors with two arguments are the the scalar product (2.45) $\eta : V \times V \to \mathbb{R}$, $(u, v) \mapsto \eta(u, v) = u \cdot v = u^m \eta_{mn} v^n$, which in an orthonormal basis has components

$$\eta_{mn} = \begin{cases} 1 & \text{if } m = n = 0 \,, \\ -1 & \text{if } m = n \in \{1, 2, 3\} \,, \\ 0 & \text{if } m \neq n \,. \end{cases}\tag{4.84}$$

or the current density j, which maps vectors $\mathbf{a}$ and $\mathbf{b}$ to the flow $j(\mathbf{a}, \mathbf{b})$ through the small parallelogram with edges $\mathbf{a}$ and $\mathbf{b}$.

The Kronecker δ is the tensor over V^* and V which maps (l, v) to $\delta(l, v) = l(v)$.

Because tensors are linear, they are completely determined by their values on basis vectors, $\eta(u, v) = \eta(e_m, e_n)u^m v^n$ or $j(\mathbf{a}, \mathbf{b}) = j(\mathbf{e}_i, \mathbf{e}_j) a^i b^j$ or $\delta(l, v) = \delta(f^m, e_n)l_m v^n$. These values define the components of the tensor, $\eta_{mn} = \eta(e_m, e_n)$ or $j_{ij} = j(\mathbf{e}_i, \mathbf{e}_j)$ and $\delta(f^m, e_n) = f^m(e_n) = \delta_n^m$ (4.80). If the tensor T depends on a dual vector $l = l_m f^m$, then to this argument there corresponds an upper index of its components, e.g. if T is a tensor over the vector spaces V and V^* and if $(u, l) \in V \times V^*$, then

$$T(u, l) = T(e_n, f^m) u^n l_m \,, \quad T(e_n, f^m) = T_n{}^m \,, \quad T(u, l) = T_n{}^m u^n l_m \,.\tag{4.85}$$

Each lower index, which enumerates tensor components, relates to a vector argument, each upper index to a dual vector argument.

Tensors can be scaled by a factor and they can be added if they are of the same type i.e. have the same arguments, e.g. the tensors T over the vector spaces V and W map $V \times W$ bilinearly to $\mathbb{R}$ and constitute a vector space. Its dual space is the tensor product $V \otimes W$. Each pair $(u, w) \in V \times W$ defines the linear map

$$u \otimes w : T \mapsto T(u, w) \tag{4.86}$$

of tensors to numbers. Because of linearity in the left argument $T(a\,u + v, w) = aT(u, w) + T(v, w)$, the tensor product is linear in the left factor and by the same argument in the right factor, i.e. for all numbers a and vectors u, v and w, z the tensor product is distributive

$$(au + v) \otimes w = a(u \otimes w) + v \otimes w \,, \ u \otimes (aw + z) = a(u \otimes w) + u \otimes z \,. \tag{4.87}$$

Because $u = e_m u^m$ and $w = e'_n w^n$ are linear combinations of a basis $e_1, e_2 \ldots e_d$ and $e'_1, e'_2 \ldots e'_{d'}$, the tensor products are linear combinations of the tensor products of the basis vectors

$$u \otimes w = e_i \otimes e'_j \, u^i \, w^j \,, \ e_i \otimes e'_j : T \mapsto T_{ij} \,. \tag{4.88}$$

The tensor product $V \otimes W$ is the space of linear combinations $v = e_i \otimes e'_j v^{ij}$. They map tensors T over V and W linearly to $v(T) = v^{ij} e_i \otimes e'_j(T) = v^{ij} T(e_i, e'_j) = v^{ij} T_{ij}$. A generic element of the tensor product does not factorize into the product of two factors but is a linear combination of products.

Under a linear transformation D of the vector space V or D^{T-1} of V^* tensors transform by right multiplication with the inverse (4.47) into e.g.

$$T' : (u, w) \mapsto T'(u, w) = T(D^{-1} u, D^T w) \,. \tag{4.89}$$

By $T'(e_m, f^n) = T(D^{-1}(e_m), D^T(f^n)) = T(e_k D_m^{-1\,k}, D_l^n \, f^l)$ the transformed tensor has components

$$T'_m{}^n = D^{T-1}{}_m{}^k \, D^n{}_l \, T_k{}^l \,, \tag{4.90}$$

i.e. the components of tensors transform like the products of the components of vectors and dual vectors with the same index position (4.83) with one transformation matrix for each index. As tensors transform linearly, $T = 0$ is a fixed point.

The Kronecker δ is a tensor which is invariant under all tensor transformations, $\delta' = \delta$.

The scalar product is invariant under Lorentz transformations by their definition, $\eta'(u, v) = \eta(\Lambda^{-1} u, \Lambda^{-1} v) = \eta(u, v)$.

By the orthogonality relation $D^T = D^{-1}$ (6.6) the contragredient rotation coincides with D. This is why one can identify a Euclidean space with its dual and need not distinguish between upper and lower indices in an orthonormal basis. However, by (6.22) the contragredient Lorentz transformation Λ^{T-1} is equivalent, but not equal, to Λ,

$$\Lambda^{T-1} = \eta \Lambda \eta^{-1} \,, \ \Lambda^{T-1}{}_m{}^n = \eta_{mk} \Lambda^k{}_l \eta^{-1\,lm} \,. \tag{4.91}$$

Therefore one has to distinguish between four-vectors and their dual vectors.

In (4.91) η is the linear map from the vector space of four-vectors, $\mathscr{V}$, to its dual $\mathscr{V}^*$ which maps each vector u to the dual vector $\eta u : v \mapsto \eta(u, v)$. The dual vector ηu has components $(\eta u)_n = u^m \eta_{mn}$ as $(\eta u)(v) = (\eta u)_n v^n = \eta(u, v)$ shows. Because the scalar product is not degenerate, the map η is invertible and defines a unique correspondence of vectors and dual vectors.

The unique correspondence allows the shorthand notation, to call ηu just u again. From the position of the index one deduces whether one deals with the components of the vector or its corresponding dual vector. The components are related by "raising and lowering the index with η"

$$u_m = \eta_{mn} u^n \text{ or } u^n = \eta^{nk} u_k , \quad (u_0, u_1, u_2, u_3) = (u^0, -u^1, -u^2, -u^3) . \quad (4.92)$$

We also use the convention to write η^{mn} rather than $\eta^{-1\,mn}$ and deduce from its upper index pair that we deal with the components of the matrix inverse of η.

Because η is symmetric, a sum of a lower with an upper index is like a seesaw: one may raise the lower index and lower the upper index, $u_n v^n = u^m \eta_{mn} v^n = u^m v_m$.

Lowering and raising indices commutes with differentiation, $\partial_k w_m = \eta_{mn} \partial_k w^n$.

η intertwines the Lorentz transformations of the vector space space and its dual, $\Lambda^{T-1}(\eta u) = \eta \Lambda \eta^{-1} \eta u = \eta(\Lambda u)$. Whether one maps a vector to its dual and transforms later or transforms first and maps to the dual afterwards is the same, the unique correspondence between vector and dual vector is preserved by their Lorentz transformations.

The flow through degenerate parallelograms with equal edge vectors vanishes,

$$0 = J(\mathbf{a} + \mathbf{b}, \mathbf{a} + \mathbf{b}) = J(\mathbf{a}, \mathbf{a}) + J(\mathbf{a}, \mathbf{b}) + J(\mathbf{b}, \mathbf{a}) + J(\mathbf{b}, \mathbf{b})$$
$$= 0 + J(\mathbf{a}, \mathbf{b}) + J(\mathbf{b}, \mathbf{a}) + 0 , \quad (4.93)$$

so J is antisymmetric, $J(\mathbf{a}, \mathbf{b}) = -J(\mathbf{b}, \mathbf{a})$. Therefore in three dimensions only the three components $J_{12} = \mathfrak{e} j^3$, $J_{31} = \mathfrak{e} j^2$ and $J_{23} = \mathfrak{e} j^1$ ($\mathfrak{e} = \mathrm{vol}(\mathbf{e}_1, \mathbf{e}_2, \mathbf{e}_3)$ is the volume of the basic parallelepiped) are independent and the current through the parallelogram $J(\mathbf{a}, \mathbf{b}) = J_{ij} a^i b^j = \sum_{i<j} J_{ij} (a^i b^j - a^j b^i)$ turns out to be the volume

$$\mathrm{vol}(\mathbf{j}, \mathbf{a}, \mathbf{b}) = \mathfrak{e} \left(j^1 (a^2 b^3 - b^3 a^2) + j^2 (a^3 b^1 - b^1 a^3) + j^3 (a^1 b^2 - b^2 a^1) \right) = \mathbf{j} \cdot (\mathbf{a} \times \mathbf{b}) \quad (4.94)$$

of the parallelepiped with edges $\mathbf{j} = \mathbf{e}_i j^i$, $\mathbf{a} = \mathbf{e}_j a^j$ and $\mathbf{b} = \mathbf{e}_k b^k$. The volume of the basic parallelepiped $\mathfrak{e}$ has to appear in this expression in order to make the volume a function of the three vector arguments which does not depend on the basis in which one chooses to represent them. The volume has the form of a triple sum, $\mathrm{vol}(\mathbf{a}, \mathbf{b}, \mathbf{c}) = \mathfrak{e} \, \varepsilon_{ijk} \, a^i \, b^j \, c^k$, where the ε-symbol has components with three indices and is antisymmetric under exchange of each pair of indices,

$$\varepsilon_{ijk} = -\varepsilon_{jik} = -\varepsilon_{kji} = -\varepsilon_{ikj} , \quad (4.95)$$

$$\varepsilon_{ijk} = \begin{cases} 1 & \text{if } i, j, k \text{ are an even permutation of } 1, 2, 3 \, , \\ -1 & \text{if } i, j, k \text{ are an odd permutation of } 1, 2, 3 \, , \\ 0 & \text{else} \, . \end{cases} \qquad (4.96)$$

With the ε-symbol we can write the vector product $(\mathbf{a} \times \mathbf{b}) = \mathbf{e}_i \, \mathfrak{e} \, g^{im} \varepsilon_{mjk} \, a^j \, b^k$, the determinant of 3×3-matrices $\det D = \varepsilon_{ijk} D^i_{\,1} \, D^j_{\,2} \, D^k_{\,3}$ and the volume of a parallelepiped with edges $\mathbf{a}, \mathbf{b}$ and $\mathbf{c}$ as $\text{vol}(\mathbf{a}, \mathbf{b}, \mathbf{c}) = \mathfrak{e} \, \varepsilon_{ijk} \, a^i \, b^j \, c^k = \mathbf{a} \cdot (\mathbf{b} \times \mathbf{c})$.

The transformation of the vector product under rotations of its factors is deduced from the definition of the determinant of a linear map D. It is the factor by which volume changes, $\text{vol}(D\mathbf{c}, D\mathbf{a}, D\mathbf{b}) = (\det D) \, \text{vol}(\mathbf{c}, \mathbf{a}, \mathbf{b})$. Moreover $\text{vol}(D\mathbf{c}, D\mathbf{a}, D\mathbf{b}) = (D\mathbf{c}) \cdot (D\mathbf{a} \times D\mathbf{b}) = (\det D) \, \mathbf{c} \cdot (\mathbf{a} \times \mathbf{b}) = (\det D) \, D\mathbf{c} \cdot D(\mathbf{a} \times \mathbf{b})$ because scalar products are invariant under rotations. This holds for all $\mathbf{c}' = D\mathbf{c}$, so $D\mathbf{a} \times D\mathbf{b} = (\det D) D(\mathbf{a} \times \mathbf{b})$.

Because $\mathbf{a} \times \mathbf{b}$ transforms under rotation of its factors as an axial vector by multiplication with $(\det D)D$ the space of axial vectors is different from the space of polar vectors which transform by multiplication with D. This is often indicated by their different units of measurement, which forbid to add them. Nevertheless one can identify directions: e.g. the x-axes are left invariant in both spaces by the same rotations. The irreducible representations of the rotation group single out invertible maps between the vector spaces, which by Schur's lemma are unique up to a factor.

Chapter 5
Electrodynamics

Abstract The electromagnetic fields and charge and current densities constitute a relativistic physical system. Changes in the charge distribution cause changes of the fields with the speed of light. Though the effects are retarded, they nevertheless approximately obey actio et reactio because the electric field of a charge in straight uniform motion points to its instantaneous, not its retarded position. Only acceleration leads to a loss of energy and momentum by radiation. The electrodynamic interactions are invariant under dilations, which is why they cannot explain the particular values of particle masses or the particular sizes of atoms.

5.1 Covariant Maxwell Equations

Electric and magnetic fields $\mathbf{E}(x)$ and $\mathbf{B}(x)$ change by the Lorentz force

$$\mathbf{F}_{\text{Lorentz}}(x, \mathbf{v}) = q\,(\mathbf{E}(x) + \mathbf{v} \times \mathbf{B}(x)) \tag{5.1}$$

the momentum $\mathbf{p}_{\text{particle}} = m\mathbf{v}/\sqrt{1 - \mathbf{v}^2}$ (3.42) and thereby the velocity $\mathbf{v}$ of a particle with mass m and charge q, which passes at time t the position $\mathbf{x} = (x^1, x^2, x^3)$

$$\frac{d\mathbf{p}_{\text{particle}}}{dt} = \mathbf{F}_{\text{Lorentz}} . \tag{5.2}$$

In (5.1) $x = (t, x^1, x^2, x^3)$ combines the cartesian coordinates of spacetime.

The electromagnetic fields are related to the charge density ρ and the current density $\mathbf{j}$ by the Maxwell equations. To stress the essential, we discuss the equations in length units with $c = 1$ and charge units with $\varepsilon_0 = 1$,

$$\operatorname{div}\mathbf{B} = 0, \qquad \operatorname{rot}\mathbf{E} + \frac{\partial}{\partial t}\mathbf{B} = 0 , \tag{5.3}$$

N. Dragon, *The Geometry of Special Relativity—a Concise Course*,
SpringerBriefs in Physics, DOI: 10.1007/978-3-642-28329-1_5,
© The Author(s) 2012

$$\operatorname{div} \mathbf{E} = \rho, \qquad \operatorname{rot} \mathbf{B} - \frac{\partial}{\partial t} \mathbf{E} = \mathbf{j} \,. \tag{5.4}$$

The divergence and the rotation of a vectorfield $\mathbf{C} = (C^1, C^2, C^3)$ denote the following linear combinations of partial derivatives

$$\operatorname{div} \mathbf{C} = \partial_1 C^1 + \partial_2 C^2 + \partial_3 C^3, \operatorname{rot} \mathbf{C} = \begin{pmatrix} \partial_2 C^3 - \partial_3 C^2 \\ \partial_3 C^1 - \partial_1 C^3 \\ \partial_1 C^2 - \partial_2 C^1 \end{pmatrix}, \tag{5.5}$$

where we employed the paper saving notation which we will use from now on,

$$\partial_0 = \frac{\partial}{\partial t}, \quad \partial_1 = \frac{\partial}{\partial x^1}, \quad \partial_2 = \frac{\partial}{\partial x^2}, \quad \partial_3 = \frac{\partial}{\partial x^3} \,. \tag{5.6}$$

Explicitly the homogeneous Maxwell equations (5.3) read

$$\begin{aligned} \partial_1 B^1 + \partial_2 B^2 + \partial_3 B^3 &= 0 \,, \\ \partial_2 E^3 - \partial_3 E^2 + \partial_0 B^1 &= 0 \,, \\ \partial_3 E^1 - \partial_1 E^3 + \partial_0 B^2 &= 0 \,, \\ \partial_1 E^2 - \partial_2 E^1 + \partial_0 B^3 &= 0 \,. \end{aligned} \tag{5.7}$$

We interpret the field strengths as the six components of an antisymmetric tensor, the field strength tensor F, which maps pairs of four-vectors v and w to the flux $F(v, w)$ of field strength, which flows through the parallelogram with edges v and w,

$$F_{mn} = -F_{nm}, m, n \in \{0, 1, 2, 3\}, F_{0i} = E^i, F_{ij} = -\varepsilon_{ijk} B^k, i, j, k \in \{1, 2, 3\}, \tag{5.8}$$

$$\begin{pmatrix} F_{00} & F_{01} & F_{02} & F_{03} \\ F_{10} & F_{11} & F_{12} & F_{13} \\ F_{20} & F_{21} & F_{22} & F_{23} \\ F_{30} & F_{31} & F_{32} & F_{33} \end{pmatrix} = \begin{pmatrix} 0 & E^1 & E^2 & E^3 \\ -E^1 & 0 & -B^3 & B^2 \\ -E^2 & B^3 & 0 & -B^1 \\ -E^3 & -B^2 & B^1 & 0 \end{pmatrix} \,. \tag{5.9}$$

Then the homogeneous Maxwell equations have the intriguing symmetric pattern

$$\begin{aligned} \partial_1 F_{23} + \partial_2 F_{31} + \partial_3 F_{12} &= 0 \,, \\ \partial_2 F_{30} + \partial_3 F_{02} + \partial_0 F_{23} &= 0 \,, \\ \partial_3 F_{01} + \partial_0 F_{13} + \partial_1 F_{30} &= 0 \,, \\ \partial_0 F_{12} + \partial_1 F_{20} + \partial_2 F_{01} &= 0 \,. \end{aligned} \tag{5.10}$$

They relate the cyclic permutations of the derivatives of the components of F,

$$\partial_l F_{mn} + \partial_m F_{nl} + \partial_n F_{lm} = 0 \,. \tag{5.11}$$

If a quantity $X_{lmn} = -X_{lnm}$ is antisymmetric in the last index pair, such as $\partial_l F_{mn}$, then its cyclic sum $Z_{lmn} = X_{lmn} + X_{mnl} + X_{nlm}$ is antisymmetric under each permutation of two indices, e.g. $Z_{mln} = X_{mln} + X_{lnm} + X_{nml} = -X_{mnl} - X_{lmn} - X_{nlm} = -Z_{lmn}$. So (5.11) does not consist of $4 \cdot 4 \cdot 4$ independent equations, as one might deduce from three indices l, m and n which range over four values. Rather l, m and n must be pairwise different in a nontrivial equation and their permutation does not lead to a new equation. Thus, (5.11) are the $4 \cdot 3 \cdot 2 / 3! = 4$ independent equations (5.10).

The inhomogeneous Maxwell equations (5.4) read explicitly

$$\begin{aligned}
\partial_1 E^1 + \partial_2 E^2 + \partial_3 E^3 &= \rho \,, \\
-\partial_0 E^1 + \partial_2 B^3 - \partial_3 B^2 &= j^1 \,, \\
-\partial_0 E^2 + \partial_3 B^1 - \partial_1 B^3 &= j^2 \,, \\
-\partial_0 E^3 + \partial_1 B^2 - \partial_2 B^1 &= j^3
\end{aligned} \tag{5.12}$$

or, in terms of the components of the field strength tensor if we insert vanishing terms like $\partial_0 F_{00}$ to emphasize the structure and denote ρ by j^0,

$$\begin{aligned}
\partial_0 F_{00} - \partial_1 F_{10} - \partial_2 F_{20} - \partial_3 F_{30} &= j^0 \,, \\
-\partial_0 F_{01} + \partial_1 F_{11} + \partial_2 F_{21} + \partial_3 F_{31} &= j^1 \,, \\
-\partial_0 F_{02} + \partial_3 F_{32} + \partial_1 F_{12} + \partial_2 F_{22} &= j^2 \,, \\
-\partial_0 F_{03} + \partial_1 F_{13} + \partial_2 F_{23} + \partial_3 F_{33} &= j^3 \,.
\end{aligned} \tag{5.13}$$

The minus signs pertain to the definition of F with raised indices (4.92)

$$F^{mn} = \eta^{mk} \eta^{nl} F_{kl}, \quad F^{00} = F_{00}, \quad F^{0i} = -F_{0i}, \quad F^{ij} = F_{ij}, i, j \in \{1, 2, 3\}, \tag{5.14}$$

$$\begin{pmatrix}
F^{00} & F^{01} & F^{02} & F^{03} \\
F^{10} & F^{11} & F^{12} & F^{13} \\
F^{20} & F^{21} & F^{22} & F^{23} \\
F^{30} & F^{31} & F^{32} & F^{33}
\end{pmatrix} = \begin{pmatrix}
0 & -E^1 & -E^2 & -E^3 \\
E^1 & 0 & -B^3 & B^2 \\
E^2 & B^3 & 0 & -B^1 \\
E^3 & -B^2 & B^1 & 0
\end{pmatrix}. \tag{5.15}$$

Then the inhomogeneous Maxwell equations have the form

$$\begin{aligned}
\partial_0 F^{00} + \partial_1 F^{10} + \partial_2 F^{20} + \partial_3 F^{30} &= j^0 \,, \\
\partial_0 F^{01} + \partial_1 F^{11} + \partial_2 F^{21} + \partial_3 F^{31} &= j^1 \,, \\
\partial_0 F^{02} + \partial_1 F^{12} + \partial_2 F^{22} + \partial_3 F^{32} &= j^2 \,, \\
\partial_0 F^{03} + \partial_1 F^{13} + \partial_2 F^{23} + \partial_3 F^{33} &= j^3 \,,
\end{aligned} \tag{5.16}$$

or in index notation

$$\partial_m F^{mn} = j^n \,. \tag{5.17}$$

Local Charge Conservation

A double sum over a symmetric and an antisymmetric pair of indices vanishes,

$$T_{rs}{}^{mn} = -T_{rs}{}^{nm} = T_{sr}{}^{mn} \Rightarrow T_{kl}{}^{kl} = 0 \,, \tag{5.18}$$

since, after permuting and relabeling the summation indices, it equals its negative,

$$T_{kl}{}^{kl} = T_{lk}{}^{kl} = -T_{lk}{}^{lk} = -T_{lm}{}^{lm} = -T_{km}{}^{km} = -T_{kl}{}^{kl} \,. \tag{5.19}$$

The field strength is antisymmetric, $F^{mn} = -F^{nm}$. Partial derivatives, if they are continuous, can be interchanged $\partial_n \partial_m = \partial_m \partial_n$, so the double sum $\partial_n \partial_m F^{mn}$ vanishes.

So, applying ∂_n to (5.17), we obtain the continuity equation $0 = \partial_n j^n$. The four-divergence $\partial_0 j^0 + \partial_1 j^1 + \partial_2 j^2 + \partial_3 j^3$ of the four-current vanishes or, in other words, the charge density decreases by the divergence of the current density,

$$\partial_n j^n = 0, \quad \partial_0 \rho + \operatorname{div} \mathbf{j} = 0 \,. \tag{5.20}$$

The continuity equation restricts conceivable sources ρ and $\mathbf{j}$ of the electromagnetic fields. Only such charge and current densities can occur in the Maxwell equations which satisfy the continuity equation and therefore local charge conservation.

Local charge conservation implies more than just charge conservation. Charge would already be conserved if it vanished in the laboratory and reappeared at the same instant on the other side of the moon. Local charge conservation, however, implies that the charge Q_V in each time independent volume V changes in the course of time only because unbalanced currents flow in and out through the boundary surface ∂V (read "boundary of V"). This follows by Gauß' theorem if one integrates the continuity equation over the volume V,

$$Q_V(t) = \int_V \mathrm{d}^3 x \, \rho(t, \mathbf{x}) \,, \tag{5.21}$$

$$\frac{\mathrm{d}}{\mathrm{d}t} Q_V(t) = \int_V \mathrm{d}^3 x \, \partial_0 \rho = -\int_V \mathrm{d}^3 x \, \operatorname{div} \mathbf{j} = -\oint_{\partial V} \mathrm{d}^2 \mathbf{f} \cdot \mathbf{j} \,. \tag{5.22}$$

In particular, a single charge cannot be produced from the vacuum.

5.2 Energy and Momentum

The electromagnetic field generates an energy density, energy currents, momentum densities and momentum currents. These quantities are the components T^{kl} of the energy-momentum tensor of the electromagnetic field

$$T^{kl} = -\left(F^k{}_n F^{ln} - \frac{1}{4}\eta^{kl} F_{mn} F^{mn}\right). \tag{5.23}$$

The energy-momentum tensor is symmetric and traceless

$$T^{kl} = T^{lk}, \quad \eta_{kl} T^{kl} = 0. \tag{5.24}$$

For each value of the index k, the T^{kl} are four components of a four-current which, in absence of electric charge and currents, is conserved,

$$\partial_l T^{lk} = -F^k{}_n j^n. \tag{5.25}$$

To verify this we calculate $-\partial_l T_k{}^l$,

$$\partial_l \left(F_{kn} F^{ln} - \frac{1}{4}\delta_k^l F_{mn} F^{mn}\right) = (\partial_l F_{kn}) F^{ln} + F_{kn} \partial_l F^{ln} - \frac{1}{2}(\partial_k F_{mn}) F^{mn}$$

$$= \frac{1}{2} (\partial_l F_{kn} - \partial_n F_{kl} - \partial_k F_{ln}) F^{ln} + F_{kn} \partial_l F^{ln} = F_{kn} j^n. \tag{5.26}$$

Here we have used the Maxwell equations (5.11, 5.17) and that the double sum with $F^{ln} = -F^{nl}$ antisymmetrizes in the indices l and n, because a double sum of a symmetric and antisymmetric index pair vanishes (5.18),

$$2F^{ln}\partial_l F_{kn} = F^{ln} (\partial_l F_{kn} - \partial_n F_{kl}) + F^{ln} (\partial_l F_{kn} + \partial_n F_{kl}) - F^{ln} (\partial_l F_{kn} - \partial_n F_{kl}). \tag{5.27}$$

If the charge and current densities vanish then the energy-momentum tensor consists of components of four conserved currents (5.25). For each k the component T^{k0} is a density whose spatial integral

$$p^k = \int d^3x \, T^{k0}(x) \tag{5.28}$$

is time-independent (5.21) if the integral over the current (T^{k1}, T^{k2}, T^{k3}) through the boundary vanishes, i.e. if no energy or momentum is radiated away.

The conserved quantities correspond, as we shall show on page 117, to the invariance of the action under temporal and spatial translations and are therefore the energy and momentum $(p^0, p^1, p^2, p^3) = (E, \mathbf{p})$. So T^{00} is the energy density and (T^{10}, T^{20}, T^{30}) is the the momentum density. Using (5.9, 5.15) we write the densities as quadratic expressions in the electric and magnetic field, $(i, j, k \in \{1, 2, 3\})$,

$$T^{00} = \frac{1}{2}\left(\mathbf{E}^2 + \mathbf{B}^2\right), \quad E = \frac{1}{2}\int d^3x \left(\mathbf{E}^2 + \mathbf{B}^2\right), \tag{5.29}$$

$$T^{i0} = \varepsilon_{ijk} E^j B^k, \qquad \mathbf{p} = \int d^3 x \mathbf{E} \times \mathbf{B}. \tag{5.30}$$

The energy density $u = \frac{1}{2} \left(\mathbf{E}^2 + \mathbf{B}^2 \right)$ is nonnegative and vanishes if and only if the field vanishes. The Poynting vector

$$\mathbf{S} = \mathbf{E} \times \mathbf{B} \tag{5.31}$$

is the momentum density of the electromagnetic field. Because the energy-momentum tensor is symmetric, $T^{0i} = T^{i0}$, the Poynting vector is also the current density of the energy.

In presence of electric charge and current density, energy and momentum of the electromagnetic field vary in time due to (5.25)

$$\partial_0 \frac{1}{2} \left(\mathbf{E}^2 + \mathbf{B}^2 \right) + \text{div} \mathbf{S} = -\mathbf{j} \cdot \mathbf{E}, \tag{5.32}$$

$$\partial_0 S^i + \partial_j T^{ij} = - \left(\rho E^i + \varepsilon_{ijk} j^j B^k \right) \tag{5.33}$$

$$\text{where} \quad T^{ij} = - \left(E^i E^j + B^i B^j - \frac{1}{2} \delta^{ij} \left(\mathbf{E}^2 + \mathbf{B}^2 \right) \right), \tag{5.34}$$

since energy and momentum can be exchanged with charged particles. If the overall momentum of particles and electromagnetic field is conserved, $\rho \mathbf{E} + \mathbf{j} \times \mathbf{B}$ must be the increase of the momentum density of the carriers of charge, and $\mathbf{j} \cdot \mathbf{E}$ is the change of their energy density.

The momentum $\mathbf{p}_{\text{particle}}$ and the energy E_{particle} of a point particle with charge q, which traverses the curve $\mathbf{x}(t)$ and has the charge density $\rho(t, \mathbf{z}) = q \, \delta^3(\mathbf{z} - \mathbf{x}(t))$ and the current density $\mathbf{j}(t, \mathbf{z}) = q \, \dot{\mathbf{x}} \, \delta^3(\mathbf{z} - \mathbf{x}(t))$, therefore change in time by

$$\frac{d\mathbf{p}_{\text{particle}}}{dt} = \mathbf{F}_{\text{Lorentz}} = q \, (\mathbf{E} + \mathbf{v} \times \mathbf{B}), \qquad \frac{dE_{\text{particle}}}{dt} = q \, \frac{d\mathbf{x}}{dt} \cdot \mathbf{E}. \tag{5.35}$$

So the Lorentz force (5.1) in the relativistic equation of motion of charged point particles states the overall conservation of energy and momentum. The energy of the particle changes by the work done by the electric field, the magnetic field does not change the particle energy.

The spatial components of the energy-momentum tensor are current densities of momentum. Through a small parallelogram with edges $\mathbf{a}$ and $\mathbf{b}$ passes the flow of momentum $F^i(\mathbf{a}, \mathbf{b}) = T^{ij}(\mathbf{a} \times \mathbf{b})^j$. If the flow is absorbed by the parallelogram, then $F^i(\mathbf{a}, \mathbf{b})$ is the force which acts on it. The isotropic part of T^{ij}, i.e. the part which is proportional to δ^{ij}, is the pressure. By (5.34) the pressure of the electromagnetic field is a third of its energy density, the energy-momentum tensor is traceless. If the radiation is not isotropic, e.g. light from the sun, then $T^{ij} n^j$ are the components of the radiation pressure exerted on an absorber with normal vector (n^1, n^2, n^3).

Uniqueness and Domain of Dependence

The electrodynamic fields depend at time $t > 0$ in position $\mathbf{x}$ only on the charges and currents and initial values at time $t = 0$ in the domain G, which is bounded by the backward light cone of $(t, \mathbf{x})$ and the spacelike initial surface I, which is cut by the backward light cone from the surface $t = 0$ [10, Chap. VI, Sect. 4],

$$G = \{(t', \mathbf{y}) : 0 \le t' \le t, |\mathbf{x} - \mathbf{y}| \le t - t'\}, \quad I = \{(0, \mathbf{y}) : |\mathbf{x} - \mathbf{y}| \le t\}. \quad (5.36)$$

For if two solutions with the same charges and currents in G coincide in their initial values of the fields $\mathbf{E}$ and $\mathbf{B}$ on the initial surface I, then the difference of both solutions satisfy the Maxwell equations with charges and currents, which vanish in G and with initial values, which vanish on I. Such a solution, however, has to vanish in G, as we are going to prove now.

The energy density u is nowhere smaller than the modulus of the energy current density $\mathbf{S}$,

$$u = \frac{1}{2}\left(\mathbf{E}^2 + \mathbf{B}^2\right) \ge |\mathbf{E} \times \mathbf{B}| = |\mathbf{S}|, \quad (5.37)$$

for $(|\mathbf{E}| - |\mathbf{B}|)^2 \ge 0$ implies $(|\mathbf{E}|^2 + |\mathbf{B}|^2) \ge 2|\mathbf{E}|\,|\mathbf{B}|$ and $|\mathbf{E}|\,|\mathbf{B}| \ge |\mathbf{E} \times \mathbf{B}|$.

Consequently for all future directed, timelike four-vectors $w = (w_0, \mathbf{w})$,

$$w_0 - |\mathbf{w}| > 0, \quad (5.38)$$

the density $w_m T^{m0}$ is nonnegative, $w_m T^{m0} \ge 0$,

$$w_m T^{m0} = w_0 u - \mathbf{w} \cdot \mathbf{S} \ge w_0 u - |\mathbf{w}||\mathbf{S}| \ge (w_0 - |\mathbf{w}|) u \ge 0 \quad (5.39)$$

and vanishes if and only if the energy density $u = (\mathbf{E}^2 + \mathbf{B}^2)/2$ and therefore all field strengths vanish, $w_m T^{m0} = 0 \Leftrightarrow \mathbf{E} = 0 = \mathbf{B}$.

Each inner point $(t', \mathbf{y})$ of the domain G lies on a spacelike surface

$$S = \{(t(\mathbf{y}), \mathbf{y}) : t(\mathbf{y}) \ge 0, \mathbf{y} \in I\}, \quad (5.40)$$

which together with I borders a domain $V \subset G$. Within V the current j^m and therefore the four-divergence $\partial_m T^{m0}$ vanish (5.25), so

$$\int_V d^4y\, \partial_m T^{m0} = 0. \quad (5.41)$$

By Gauß' theorem the integral over $d\omega = dt\, d^3y\, \partial_m T^{m0}$ over the volume V equals the integral over $\omega = d^3y\, T^{00} - \frac{1}{2} dt dy^i dy^j\, \varepsilon_{ijk}\, T^{k0}$ over the boundary I and S. But I does not contribute, because the initial values vanish, therefore

Fig. 5.1 Domain of dependence G

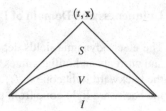

$$\int_S d^3y \left(T^{00} - \partial_i t \, T^{i0} \right) = 0 \, . \tag{5.42}$$

The dual four-vector $w = (1, -\partial_1 t, -\partial_2 t, -\partial_3 t)$ is future directed and spacelike everywhere because S is spacelike. Therefore the integrand $w_m \, T^{m0}$ is nonnegative and the integral vanishes only if $\mathbf{E}$ and $\mathbf{B}$ vanish everywhere on S. So both fields vanish in each point $(t', \mathbf{y})$ of the interior of G. Because the fields are continuous, they vanish in all of G.

Each solution of the Maxwell equations is uniquely determined by the charge and current densities and the initial values of $\mathbf{E}$ and $\mathbf{B}$ on a spacelike surface I. The fields at $(t, \mathbf{x})$ with $t > 0$ depend only on the initial values of that part of the surface I, which lies in the backward light cone of $(t, \mathbf{x})$ and on the charge and current densities in the domain G, which is bounded by the backward light cone and the initial surface (Fig. 5.1).

For $t < 0$ the corresponding results hold for the forward light cone.

Localized changes in the initial data and in the charges and currents cause changes in the fields at most with the velocity of light. Nothing outruns light.

5.3 The Electrodynamic Potentials

The electric field $\mathbf{E}$ of a spherically symmetric, static charge distribution follows from the obvious ansatz that $\mathbf{E}$ is time-independent and radial[1]

$$\mathbf{E}(\mathbf{x}) = \frac{\mathbf{x}}{|\mathbf{x}|} E(|\mathbf{x}|) \, . \tag{5.43}$$

The modulus of the $\mathbf{E}$-field is determined by integrating the Maxwell equation $\operatorname{div} \mathbf{E} = \rho$ (5.4) over a ball B_r with radius r. On the right-hand side one obtains the charge $Q(r)$ in the ball. By Gauß' theorem the left-hand side equals the integral over the boundary of the ball, the sphere ∂B_r,

---

[1] Unfortunately, "energy" and "electric field" begin with the same letter. We denote the absolute value of the electric field and the energy by E and let the context resolve the ambiguity.

$$Q(r) = \int_{B_r} d^3x \, \text{div } \mathbf{E}(\mathbf{x}) = \oint_{\partial B_r} d\mathbf{f} \cdot \mathbf{E}(\mathbf{x}) \,. \tag{5.44}$$

On the sphere the outward normal $d\mathbf{f}$ is parallel to the electric field $\mathbf{E}$, so the scalar product $d\mathbf{f} \cdot \mathbf{E}$ equals the product of the absolute values. The modulus of the electric field is constant on the sphere and can be extracted from the integral, which gives the size $4\pi r^2$ of the sphere. We obtain $Q(r) = 4\pi r^2 E(r)$,

$$E(r) = \frac{1}{4\pi} \frac{Q(r)}{r^2} \,. \tag{5.45}$$

For a spherically symmetric charge distribution only those charges influence a test charge at the point $\mathbf{x}$ which are within the sphere of radius $|\mathbf{x}|$.

The electrostatic force $\mathbf{F} = q\mathbf{E}$ repels charges q with the same sign as $Q(r)$.

Inside a homogeneously charged sphere with radius R the ratio of $Q(r)$ to the total charge Q equals the ratio of the volume $\frac{4}{3}\pi r^3$ to the total volume $\frac{4}{3}\pi R^3$, so one has $Q(r) = Qr^3/R^3$ and the electric field of a homogeneously charged sphere is

$$E(r) = \frac{Q}{4\pi} \cdot \begin{cases} \dfrac{r}{R^3} & \text{if } r < R \\[2mm] \dfrac{1}{r^2} & \text{if } r \geq R \end{cases} \,. \tag{5.46}$$

A test particle in a homogeneously charged ball with opposite charge is subject to the same force, increasing linearly with the distance, as a spherically symmetric harmonic oscillator. It orbits ellipses around the center different from Kepler ellipses around a focal point.

The electrostatic field can be written as the gradient of a potential $\phi(\mathbf{x})$ and therefore also satisfies the remaining Maxwell equations with $\mathbf{B} = 0$ and $\mathbf{j} = 0$,

$$\mathbf{E} = -\,\text{grad}\,\phi, \quad \phi(\mathbf{x}) = \frac{Q}{4\pi} \cdot \begin{cases} -\dfrac{\mathbf{x}^2}{2R^3} + \dfrac{3}{2R} & \text{if } |\mathbf{x}| < R \\[2mm] \dfrac{1}{|\mathbf{x}|} & \text{if } |\mathbf{x}| \geq R \end{cases} \,. \tag{5.47}$$

Poisson Equation

The static potential outside of a point particle at the origin with charge q is the Coulomb potential $\phi : \mathbf{y} \mapsto q/(4\pi \, |\mathbf{y}|)$. If the particle is at $\mathbf{x}$, the corresponding potential $\phi(\mathbf{y}) = q/(4\pi \, |\mathbf{y} - \mathbf{x}|)$ is the shifted potential, i.e. the Maxwell equations are translation-invariant. The potential of several point charges is the sum of the potentials of the individual charges, because the Maxwell equations are linear,

$$\phi(\mathbf{y}) = \frac{1}{4\pi} \sum_i \frac{q_i}{|\mathbf{y} - \mathbf{x}_i|} . \tag{5.48}$$

For a continuous, static charge distribution ρ the potential becomes the integral

$$\phi(\mathbf{y}) = \frac{1}{4\pi} \int d^3x \, \frac{\rho(\mathbf{x})}{|\mathbf{x} - \mathbf{y}|} . \tag{5.49}$$

It satisfies the the Poisson equation, $- \operatorname{div} \operatorname{grad} \phi = \rho$, i.e. $\Delta\phi = -\rho$, where

$$\Delta = \operatorname{div} \operatorname{grad} = \partial_1{}^2 + \partial_2{}^2 + \partial_3{}^2 \tag{5.50}$$

is the Laplace operator.

To show that (5.49) solves the Poisson equation, we cut out of the integration volume V a ball $B_{\varepsilon,\mathbf{y}}$ around $\mathbf{y}$ with radius ε

$$B_{\varepsilon,\mathbf{y}} = \{\mathbf{x} : |\mathbf{x} - \mathbf{y}| \le |\varepsilon|\} \tag{5.51}$$

and consider the integral over the remaining volume $V_\varepsilon = V - B_{\varepsilon,\mathbf{y}}$ for $\varepsilon \to 0$. Within V_ε the function $1/|\mathbf{x} - \mathbf{y}|$ is differentiable and satisfies $\Delta\frac{1}{|\mathbf{x}-\mathbf{y}|} = 0$. So in V_ε one has

$$I_\varepsilon(\mathbf{y}) = \int_{V_\varepsilon} d^3x \, \frac{1}{|\mathbf{x} - \mathbf{y}|} \Delta\phi(\mathbf{x}) = \int_{V_\varepsilon} d^3x \left(\frac{1}{|\mathbf{x} - \mathbf{y}|} \Delta\phi(\mathbf{x}) - \left(\Delta \frac{1}{|\mathbf{x} - \mathbf{y}|}\right) \phi(\mathbf{x}) \right).$$
$$\tag{5.52}$$

The integrand is a sum of derivative terms

$$\frac{1}{|\mathbf{x} - \mathbf{y}|} \Delta\phi(\mathbf{x}) - \left(\Delta \frac{1}{|\mathbf{x} - \mathbf{y}|}\right) \phi(\mathbf{x}) = \partial_i \left(\frac{1}{|\mathbf{x} - \mathbf{y}|} \partial_i \phi(\mathbf{x}) - \left(\partial_i \frac{1}{|\mathbf{x} - \mathbf{y}|}\right) \phi(\mathbf{x}) \right).$$

Therefore, according to Gauß' theorem

$$I_\varepsilon(\mathbf{y}) = \int_{\partial V_\varepsilon} d^2\mathbf{f} \cdot \left(\frac{1}{|\mathbf{x} - \mathbf{y}|} \operatorname{grad} \phi(\mathbf{x}) - \left(\operatorname{grad} \frac{1}{|\mathbf{x} - \mathbf{y}|}\right) \phi(\mathbf{x}) \right)$$

$$= \int_{\partial V_\varepsilon} d^2\mathbf{f} \cdot \left(\frac{1}{|\mathbf{x} - \mathbf{y}|} \operatorname{grad} \phi(\mathbf{x}) + \frac{\mathbf{x} - \mathbf{y}}{|\mathbf{x} - \mathbf{y}|^3} \phi(\mathbf{x}) \right). \tag{5.53}$$

The boundary of V_ε consists of the boundary of V and the sphere $\partial B_{\varepsilon,\mathbf{y}}$. Observe, that in the surface integral (5.53) the normal vector $d^2\mathbf{f} = d^2 f \, \mathbf{n}$ points out of V_ε into the ball $B_{\varepsilon,\mathbf{y}}$. On the sphere $\partial B_{\varepsilon,\mathbf{y}}$ one has

$$\frac{1}{|\mathbf{x}-\mathbf{y}|} = \frac{1}{\varepsilon}, \qquad \frac{\mathbf{x}-\mathbf{y}}{|\mathbf{x}-\mathbf{y}|^3} = -\frac{1}{\varepsilon^2}\,\mathbf{n}\,. \tag{5.54}$$

The integral over the first term vanishes in the limit $\varepsilon \to 0$,

$$\int_{\partial B_{\varepsilon,\mathbf{y}}} d^2\mathbf{f} \cdot \frac{1}{|\mathbf{x}-\mathbf{y}|}\,\mathrm{grad}\,\phi(\mathbf{x}) = \frac{1}{\varepsilon}\int_{\partial B_{\varepsilon,\mathbf{y}}} d^2\mathbf{f} \cdot \mathrm{grad}\,\phi(\mathbf{x}) \longrightarrow 0\,, \tag{5.55}$$

because by the mean value theorem for integration it equals a value of $(\mathbf{n} \cdot \mathrm{grad}\,\phi)$ at some point of the sphere times its size $4\pi\varepsilon^2$ divided by ε.

The scalar product in the second term of the integrand is minus the product of the absolute values. By the mean value theorem on some point $\mathbf{z}$ of the sphere $\partial B_{\varepsilon,\mathbf{y}}$

$$\int_{\partial B_{\varepsilon,\mathbf{y}}} d^2\mathbf{f}\,\frac{\mathbf{x}-\mathbf{y}}{|\mathbf{x}-\mathbf{y}|^3}\,\phi(\mathbf{x}) = -\frac{1}{\varepsilon^2}\int_{\partial B_{\varepsilon,\mathbf{y}}} d^2 f\,\phi(\mathbf{x}) = -\frac{4\pi\varepsilon^2}{\varepsilon^2}\,\phi(\mathbf{z})\,. \tag{5.56}$$

As $\mathbf{z}$ tends to the center $\mathbf{y}$ of $B_{\varepsilon,\mathbf{y}}$ in the limit of vanishing ε the integral over $\partial B_{\varepsilon,\mathbf{y}}$ tends to $-4\pi\phi(\mathbf{y})$.

Altogether we obtain

$$\int_V d^3x\,\frac{1}{|\mathbf{x}-\mathbf{y}|}\,\Delta\phi(\mathbf{x}) = \int_{\partial V} d^2\mathbf{f} \cdot \left(\frac{1}{|\mathbf{x}-\mathbf{y}|}\,\mathrm{grad}\,\phi(\mathbf{x}) + \frac{\mathbf{x}-\mathbf{y}}{|\mathbf{x}-\mathbf{y}|^3}\,\phi(\mathbf{x})\right) - 4\pi\phi(\mathbf{y})$$

$$\tag{5.57}$$

or, solved for $\phi(\mathbf{y})$,

$$\phi(\mathbf{y}) = -\frac{1}{4\pi}\int_V d^3x\,\frac{\Delta\phi(\mathbf{x})}{|\mathbf{x}-\mathbf{y}|} + \frac{1}{4\pi}\int_{\partial V} d^2\mathbf{f} \cdot \left(\frac{1}{|\mathbf{x}-\mathbf{y}|}\,\mathrm{grad}\,\phi(\mathbf{x}) + \frac{\mathbf{x}-\mathbf{y}}{|\mathbf{x}-\mathbf{y}|^3}\,\phi(\mathbf{x})\right)\,. \tag{5.58}$$

Each function ϕ with continuous second derivatives in $V \subset \mathbb{R}^3$, which is continuous on the closure of V, is determined by $\Delta\phi$ and its boundary values on ∂V.

The boundary values of the electrostatic potential of spatially bounded charge distributions can be chosen to vanish for $V = \mathbb{R}^3$, i.e. (5.49) solves the Poisson equation for insular charge distributions with vanishing boundary values.

Harmonic Functions

Functions ϕ, which in a domain V solve the Laplace equation,

$$\Delta\phi = 0\,, \tag{5.59}$$

e.g. the electrostatic potential in a charge free region, are called harmonic in V.

The surface integral over the normal derivative of harmonic functions over the boundary surface of V yields the charge within V and therefore vanishes

$$\int\limits_{\partial V} d^2\mathbf{f} \cdot \text{grad}\,\phi = \int\limits_V d^3x\,\partial_i\partial_i\phi = 0 \,. \qquad (5.60)$$

The representation (5.58) of functions with continuous second derivatives in V also holds for each ball $B_{R,\mathbf{y}} \subset V$ (5.51) around some inner point $\mathbf{y}$ of V. Because ϕ is harmonic in V, the volume integral over $B_{R,\mathbf{y}}$ and the surface integral over the normal derivative vanish, because the factor $1/R$ is constant on the boundary. So ϕ equals at $\mathbf{y}$ its mean value $M_{R,\mathbf{y}}[\phi]$ on the sphere $\partial B_{R,\mathbf{y}}$, $\phi(y) = M_{R,y}[\phi]$,

$$M_{R,\mathbf{y}}[\phi] = \frac{1}{4\pi} \int\limits_{\cos\theta=-1}^{\cos\theta=+1} d\cos\theta \int\limits_0^{2\pi} d\varphi\,\phi\,(\mathbf{y} + R\mathbf{n}(\theta,\varphi)) \,, \qquad (5.61)$$

where $\mathbf{n}$ denotes the unit vector with angles θ, φ (2.34),

$$\mathbf{n}(\theta,\varphi) = (\sin\theta\cos\varphi, \sin\theta\sin\varphi, \cos\theta) \,. \qquad (5.62)$$

As each harmonic function equals its mean value on surrounding spheres, it becomes minimal or maximal on the boundary of the domain V, in which it is harmonic. In particular, as an electrostatic potential does not have a local minimum in a charge free region, there is no electrostatic trap for charged particles.

Each solution ϕ of the Poisson equation $\Delta\phi = -\rho$ is uniquely determined by its boundary values and the source ρ, because the difference of two solutions with the same ρ and the same boundary values is a solution of the Laplace equation, which vanishes on the boundary ∂V and therefore in the enclosed volume V.

The potential on a conducting surface ∂V becomes constant after all currents have faded away. If the surface encloses a charge free region V, then the potential is also constant in V because its values are between the minimal and maximal values at ∂V. Therefore the electric field strength vanishes in each Faraday cage.

If the normal derivative $n^i\partial_i\varphi$ of a harmonic function vanishes on ∂V

$$0 = -\int\limits_V d^3x\,\phi\,\Delta\varphi = \int\limits_V d^3x\,\partial_i\varphi\,\partial_i\phi - \int\limits_{\partial V} d^2f\,n^i\,\phi\,\partial_i\varphi = \int\limits_V d^3x\,\sum(\partial_i\phi)^2 \,,$$

$$(5.63)$$

then the gradient $\partial_i\phi$ vanishes in V and ϕ is constant.

The normal derivative on the boundary determines the solution up to a constant, because the difference of two solutions with the same normal derivative is a harmonic function with vanishing normal derivative and therefore constant.

As ϕ^2 and $(\partial_i\phi)^2$ are nonnegative, (5.63) shows, that in regions without boundary the Laplace operator Δ does not have positive eigenvalues, $\Delta\phi = \lambda\phi \Rightarrow \lambda \leq 0$.

Gauge Transformation of the Four-Potential

The homogeneous Maxwell equations (5.11) imply that the field strengths $F_{kl}(x)$ are the antisymmetrized derivatives of four potential functions $A = (A_0, A_1, A_2, A_3)$,

$$F_{kl}(x) = \partial_k A_l(x) - \partial_l A_k(x), k, l \in \{0, 1, 2, 3\} \ . \tag{5.64}$$

Field strengths of this form solve the homogeneous Maxwell equations, for (5.11) is totally antisymmetric under permutations and the order of partial derivatives can be interchanged $\partial_k \partial_l A_m - \partial_l \partial_k A_m = 0$.

Conversely, one verifies by differentiation that the antisymmetrized derivatives of the following four-potential (which is defined in star-shaped regions, which contain with each point x also the connecting line to the origin)

$$A_l(x) = \int\limits_0^1 d\lambda\, \lambda x^m F_{ml}(\lambda x) \tag{5.65}$$

yield the field strengths, if they satisfy the homogeneous Maxwell equations (5.11),

$$\partial_k A_l - \partial_l A_k = \int\limits_0^1 d\lambda \left(\lambda \delta_k^m F_{ml}(\lambda x) + \lambda x^m \cdot \lambda (\partial_k F_{ml})_{|(\lambda x)} \right) - k \leftrightarrow l$$

$$= \int\limits_0^1 d\lambda\, 2\lambda F_{kl}(\lambda x) + \lambda^2 x^m (\partial_k F_{ml} - \partial_l F_{mk})_{|(\lambda x)}$$

$$\overset{5.11}{=} \int\limits_0^1 d\lambda \left(2\lambda F_{kl}(\lambda x) + \lambda^2 x^m \partial_m F_{kl}(\lambda x) \right)$$

$$= \int\limits_0^1 d\lambda \frac{\partial}{\partial \lambda} \left(\lambda^2 F_{kl}(\lambda x) \right) = \lambda^2 F_{kl}(\lambda x) \Big|_{\lambda=0}^{\lambda=1} = F_{kl}(x) \ . \tag{5.66}$$

The component A_0 is called scalar potential ϕ. The spatial components are combined to the vector potential $\mathbf{A} = (-A_1, -A_2, -A_3)$. Then (5.64) states for the magnetic and the electric field (5.9)

$$\mathbf{B} = \mathrm{rot}\mathbf{A}, \quad \mathbf{E} = -\,\mathrm{grad}\,\phi - \partial_0 \mathbf{A} \ . \tag{5.67}$$

Because of $\partial_k \partial_l \chi = \partial_l \partial_k \chi$ and (5.64) the field strengths do not change if one adds to the four-potential A_l the derivative of a function χ

$$A'_l = A_l + \partial_l \chi .\tag{5.68}$$

This change (5.68) of the four-potential is called a gauge transformation. It changes the scalar potential ϕ and the vector potential $\mathbf{A}$ into

$$\phi' = \phi + \partial_0 \chi , \quad \mathbf{A}' = \mathbf{A} - \operatorname{grad} \chi .\tag{5.69}$$

Four-potentials related by gauge transformations cannot be distinguished from each other by any physical effect.

Relativistic quantum field theory provides a deep reason for gauge invariance [11]. A quantized four-potential A creates among others states of negative norm which participate in physical processes and lead to logical contradictions of the theory unless all physical processes are gauge invariant.

5.4 Wave Equation

If one raises the indices (4.92, 5.14) of $F_{kl} = \partial_k A_l - \partial_l A_k$

$$\partial^m = \eta^{mk} \partial_k , \quad A^n = \eta^{nl} A_l ,\tag{5.70}$$

and inserts $F^{mn} = \eta^{mk} \eta^{nl} F_{kl}$ into the inhomogeneous Maxwell equations (5.17), one obtains

$$\partial_m \partial^m A^n - \partial_m \partial^n A^m = j^n .\tag{5.71}$$

In the second term the derivatives can be interchanged $\partial_m \partial^n A^m = \partial^n \partial_m A^m$ and by a choice of the gauge one can satisfy the Lorenz condition,[2]

$$\partial_m A^m = 0 ,\tag{5.72}$$

which makes the second term vanish. Then each component of the four-potential satisfies separately an inhomogeneous wave equation

$$\Box A^n = j^n .\tag{5.73}$$

The differential operator in the wave equation (pronounced "box")

$$\Box = \partial_m \partial^m = \eta^{mk} \partial_m \partial_k = \partial_0^2 - \partial_1^2 - \partial_2^2 - \partial_3^2\tag{5.74}$$

is called the wave operator or d'Alembert operator.

If $\partial_m A'^m = f$ does not vanish, then the Lorenz condition (5.72) holds for the gauge transformed four-potential $A = A' - \partial \chi$ if χ satisfies $\Box \chi = f$.

---

[2] This gauge condition is due to Ludvig Valentin Lorenz, not Hendrik Antoon Lorentz [16, 20].

Uniqueness and Domain of Dependence

The solutions u of the inhomogeneous wave equation

$$\Box u - g \tag{5.75}$$

is uniquely determined by the source g and by the initial values of u and $\partial_0 u$ at $t = 0$ and depends at $(t, \mathbf{x})$ only on the source in the domain G and the initial values in I (5.36). This is deduced just as the corresponding result for the electromagnetic fields with the arguments of page 97 from the energy-momentum tensor, which corresponds to the wave equation,

$$T_{mn} = \partial_m u \, \partial_n u - \frac{1}{2} \eta_{mn} \partial_r u \, \partial^r u \ . \tag{5.76}$$

It is conserved $\partial_m T^{mn} = 0$, if u satisfies the homogeneous wave equation $\Box u = 0$, and is positive definite for each future directed timelike vector $w = (1, w_1, w_2, w_3)$,

$$
\begin{aligned}
2 w_m T^{m0} &= (\partial_0 u)^2 + \partial_i u \, \partial_i u - 2 w_i \, \partial_i u \, \partial_0 u \\
&= (\partial_0 u - w_i \, \partial_i u)^2 + (\partial_i u \, \partial_i u) - (w_i \partial_i u)^2 \\
&\geq (\partial_0 u - w_i \, \partial_i u)^2 + (\partial_i u \, \partial_i u)(1 - w_j w_j) \geq 0 \ .
\end{aligned} \tag{5.77}
$$

The sum $w_m T^{m0}$ vanishes only, if all derivatives $\partial_m u$ vanish. By the arguments of page 97 the difference of two solutions with the same source and with the same initial values is constant on each spacelike surface S within the domain G. It vanishes, because S and I have common points. So it vanishes in G.

Plane Waves

All terms in the homogeneous wave equation are derivative terms of the same order, two, and the quadratic form $k^2 = k \cdot k = \eta^{mn} k_m k_n = k_0{}^2 - k_1{}^2 - k_2{}^2 - k_3{}^2$ which corresponds to the wave operator $\Box = \eta^{mn} \partial_m \partial_n$ allows for real, lightlike vectors $k \neq 0$, $k^2 = 0$. Therefore the wave equation $\Box u = 0$ has the remarkable property to allow plane waves $u(x) = f(k \cdot x)$ of arbitrary form f (as long as f has continuous second derivatives)

$$\Box f(k \cdot x) = \eta^{mn} k_m k_n f''(k \cdot x) = k^2 f''(k \cdot x) = 0 \ . \tag{5.78}$$

Plane waves of the form $u(t, \mathbf{x}) = f(t - \mathbf{n}\,\mathbf{x})$ with arbitrary, smooth f move with the speed of light in direction $\mathbf{n}$, $\mathbf{n}^2 = 1$, and solve the homogeneous wave equation of the four-dimensional spacetime.

The plane wave deserves its name: it is constant on planes orthogonal to $\mathbf{n}$,

$$u(t, \mathbf{x} + \mathbf{x}_0) = f(t - \mathbf{n}\,(\mathbf{x} + \mathbf{x}_0)) = f(t - \mathbf{n}\,\mathbf{x}) = u(t, \mathbf{x}) \quad \text{if} \quad \mathbf{n}\,\mathbf{x}_0 = 0 \ . \tag{5.79}$$

It is also constant along the worldlines of photons, which move with the speed of light in the direction **n**,

$$u(t, \mathbf{x}(0) + t\,\mathbf{n}) = f(t - \mathbf{n}\,(\mathbf{x}(0) + t\,\mathbf{n})) = f(-\mathbf{n}\,\mathbf{x}(0)) = u(0, \mathbf{x}(0)) \,. \quad (5.80)$$

A real, plane monochromatic wave u is more particularly of the form

$$u(t, \mathbf{x}) = \Re\, a\, e^{-ik\cdot x} = \Re\, a\, e^{i(\mathbf{k}\mathbf{x}-\omega t)} = A\cos(\mathbf{k}\,\mathbf{x} - \omega\, t + \alpha)\,, \quad (5.81)$$

where $a = Ae^{i\alpha}$ is constant. The circular frequency $\omega = k^0$ is determined up to its sign by $0 = k^2 = \omega^2 - \mathbf{k}^2$ as the absolute value of the wave vector $\mathbf{k}$. We choose ω to be positive,

$$\omega = k^0 = \sqrt{\mathbf{k}^2}\,, \quad (5.82)$$

and write an exponential factor with negative k^0 as $e^{ik\cdot x}$.

A superposition of monochromatic plane waves with different wave vectors is called a wave packet

$$u(t, \mathbf{x}) = \int \tilde{d}k\, \left(e^{i(\mathbf{k}\mathbf{x}-k^0 t)}\, a^*(\mathbf{k}) + e^{-i(\mathbf{k}\mathbf{x}-k^0 t)}\, a(\mathbf{k})\right),\quad k^0 = \sqrt{\mathbf{k}^2}\,, \quad (5.83)$$

$$\tilde{d}k = \frac{d^3k}{(2\pi)^3\, 2\,k^0}\,. \quad (5.84)$$

It solves the wave equation, $\Box u = 0$, with initial values $u_{|t=0} = \psi$ and $(\partial_0 u)_{|t=0} = \phi$,

$$\psi(\mathbf{x}) = \int \tilde{d}k\, e^{i\mathbf{k}\mathbf{x}}\,(a^*(\mathbf{k}) + a(-\mathbf{k})) = \int \frac{d^3k}{(2\pi)^3}\, e^{i\mathbf{k}\mathbf{x}}\,\tilde{\psi}(\mathbf{k}),$$

$$\phi(\mathbf{x}) = \int \tilde{d}k\, e^{i\mathbf{k}\mathbf{x}}\,(-ik^0)\,(a^*(\mathbf{k}) - a(-\mathbf{k})) = \int \frac{d^3k}{(2\pi)^3}\, e^{i\mathbf{k}\mathbf{x}}\,\tilde{\phi}(\mathbf{k})\,, \quad (5.85)$$

which determine by their Fourier amplitudes $\tilde{\psi}$ and $\tilde{\phi}$ the amplitudes a and a^*

$$a^*(\mathbf{k}) = k^0\,\tilde{\psi}(\mathbf{k}) + i\,\tilde{\phi}(\mathbf{k})\,,\quad a(\mathbf{k}) = k^0\,\tilde{\psi}(-\mathbf{k}) - i\,\tilde{\phi}(-\mathbf{k})\,. \quad (5.86)$$

Huygens' Principle

It seems physically plausible that the wave ϕ at the position $\mathbf{x}$ at time $t > 0$ should be a superposition of the earlier wave at $t = 0$ from all the places $\mathbf{x} + t\mathbf{n}$, $\mathbf{n}^2 = 1$, from which one can reach $\mathbf{x}$ with the speed of light in the run time t. To clarify this assumption we consider the mean value (5.61) $M_{t,\mathbf{x}}[\phi]$ of a function ϕ on spheres $\partial B_{t,\mathbf{x}}$ around the point $\mathbf{x}$ with radius t. The mean value (5.61) is also defined for negative t, because the mean value over all directions coincides with the mean value over all opposite directions, i.e. $M_{t,\mathbf{x}}[\phi]$ is an even function of t with a continuation to $t = 0$,

$$M_{t,\mathbf{x}}[\phi] = M_{-t,\mathbf{x}}[\phi], \quad M_{0,\mathbf{x}}[\phi] = \phi(\mathbf{x}). \tag{5.87}$$

It has continuous second derivatives if ϕ has.

The derivative with respect to t yields a surface integral over the normal derivative

$$\partial_0 M_{t,\mathbf{x}}[\phi] = \frac{1}{4\pi} \int d\cos\theta\, d\varphi\, n^i\, \partial_i \phi(\mathbf{x} + t\, \mathbf{n}) = \frac{1}{4\pi t^2} \int_{\partial B_{t,\mathbf{x}}} d^2 f\, n^i\, \partial_i \phi(\mathbf{y}). \tag{5.88}$$

For $t > 0$ the vector $\mathbf{n}$ is the outward directed normal vector. By Gauß' theorem the surface integral equals the volume integral over the divergence

$$\partial_0 M_{t,\mathbf{x}}[\phi] = \frac{1}{4\pi t^2} \int_0^t dr\, r^2 \int d\cos\theta\, d\varphi\, \Delta\phi. \tag{5.89}$$

This holds also for negative t, because then $\mathbf{n}$ is the inward directed normal vector at $\mathbf{y} = \mathbf{x} + t\, \mathbf{n}$.

If we differentiate again with respect to t and if the derivative acts by the product rule on $1/t^2$, then we obtain a contribution which coincides with $\partial_0 M$ up to a factor $-2/t$. The derivative with respect to the upper boundary t of the integral yields the integrand, which is the mean value of the derivative $\Delta\phi$ and which equals the derivative $\Delta M_{t,\mathbf{x}}[\phi]$ of the mean value,

$$\partial_0{}^2 M_{t,\mathbf{x}}[\phi] = -\frac{2}{t}\, \partial_0 M_{t,\mathbf{x}}[\phi] + \Delta M_{t,\mathbf{x}}[\phi]. \tag{5.90}$$

So the mean value satisfies Darboux's differential equation,

$$\left(\partial_0{}^2 + \frac{2}{t}\, \partial_0 - \Delta\right) M_{t,\mathbf{x}}[\phi] = 0, \tag{5.91}$$

consequently $u : (t, \mathbf{x}) \mapsto t\, M_{t,\mathbf{x}}[\phi]$ solves the homogeneous wave equation $\Box u = 0$. This holds also for $t = 0$, as the second derivatives of $t\, M_{t,\mathbf{x}}[\phi]$ are continuous.

The solution $t\, M_{t,\mathbf{x}}[\phi]$ vanishes initially at $t = 0$, its time derivative at $t = 0$ in position $\mathbf{x}$ has the value $\phi(\mathbf{x})$.

The coefficients η^{nm} of the wave equation $\eta^{mn} \partial_m \partial_n u = 0$ are constant (5.74). Therefore the derivative $\partial_0(t\, M_{t,\mathbf{x}}[\psi])$ of a solution of the wave equation is also a solution. $\partial_0(t\, M_{t,\mathbf{x}}[\psi])$ has vanishing time derivative at $t = 0$, because it is an even function of t, and assumes at $t = 0$ in position $\mathbf{x}$ the value $\psi(\mathbf{x})$. Therefore

$$u(t, \mathbf{x}) = t\, M_{t,\mathbf{x}}[\phi] + \partial_0\left(t\, M_{t,\mathbf{x}}[\psi]\right) \tag{5.92}$$

solves in $1+3$-dimensions the wave equation $\Box u = 0$ with the initial values

$$u(0, \mathbf{x}) = \psi(\mathbf{x}), \quad \partial_0 u(0, \mathbf{x}) = \phi(\mathbf{x}) \,. \tag{5.93}$$

Each solution of the wave equation is uniquely determined by its initial values (page 104). Accordingly (5.92) is *the* solution with the initial values ψ and ϕ. As it depends on the mean of the initial values it changes little if the initial values change little. The initial value problem of the wave equation is well-posed.

The solution (5.92) satisfies Huygens' principle: at time t in position $\mathbf{x}$ it is composed additively from the initial values at time $t = 0$ from all points $\mathbf{y}$, from which one can reach $\mathbf{x}$ in time t with the speed of light. This principle does not hold in even space dimensions and in $d = 1$ dimensional space. There also the initial values from nearer places contribute to the solution. In $d = 1, 2, 4, \ldots$ space dimensions the initial values reverberate.

Retarded Potential

To get an idea of the solution of the inhomogeneous wave equation, we integrate

$$\partial_0{}^2 u - \Delta u = \hat{g} \tag{5.94}$$

over the short time interval $\tau < t < \tau + \varepsilon$. The initial values of u at time τ are taken to vanish, $u(\tau, \mathbf{x}) = 0$, $\partial_t u(\tau, \mathbf{x}) = 0$. On the left hand side we obtain $\partial_0 u(\tau + \varepsilon)$ up to terms which, because of the initial conditions, are of second order in ε, the result of the right hand side is simply given a name,

$$\partial_0 u(\tau + \varepsilon) = \int_\tau^{\tau+\varepsilon} \mathrm{d}t \left(\partial_0{}^2 u - \Delta u \right) = \int_\tau^{\tau+\varepsilon} \mathrm{d}t \, \hat{g}(t, \mathbf{x}) = g(\tau + \varepsilon, \mathbf{x}) \,. \tag{5.95}$$

If the inhomogeneity $\hat{g}$ differs from zero only for this short interval, then subsequently for $t > \tau$ the corresponding solution φ_τ solves $\Box \varphi_\tau = 0$ with initial values

$$\varphi_\tau(\tau, \mathbf{x}) = 0, \quad \partial_0 \varphi_\tau(t, \mathbf{x})|_{t=\tau} = g_\tau(\mathbf{x}), \quad g_\tau(\mathbf{x}) = g(\tau, \mathbf{x}) \,, \tag{5.96}$$

up to terms which vanish with ε. φ_τ is given by (5.92) translated in time by τ

$$\varphi_\tau(t, \mathbf{x}) = (t - \tau) \, M_{(t-\tau), \mathbf{x}}[g_\tau] \,. \tag{5.97}$$

If the inhomogeneity g acts longer, then we compose the solution from solutions, which are generated by short sources: if u_1 is a solution of a linear inhomogeneous equation $L u_1 = g_1$ with source g_1 and if u_2 is a solution with source g_2, then the sum $u = u_1 + u_2$ is a solution with the sum of the sources, $L(u_1 + u_2) = (g_1 + g_2)$.

To make these considerations precise, we consider the integral

$$u(t, \mathbf{x}) = \int_0^t d\tau \, \varphi_\tau(t, \mathbf{x}) \,. \tag{5.98}$$

It vanishes initially at $t = 0$, as the domain of integration vanishes then.

Its time derivative differentiates with respect to the upper boundary of the integral, where it yields the integrand, and with respect to the first argument of φ_τ,

$$\partial_0 \, u(t, \mathbf{x}) = \varphi_t(t, \mathbf{x}) + \int_0^t d\tau \, \partial_0 \, \varphi_\tau(t, \mathbf{x}) = \int_0^t d\tau \, \partial_0 \, \varphi_\tau(t, \mathbf{x}) \,. \tag{5.99}$$

The time derivative vanishes at $t = 0$ together with the domain of integration.

The second time derivative is

$$\partial_0^2 \, u(t, \mathbf{x}) = \partial_0 \, \varphi_\tau(t, \mathbf{x})|_{\tau=t} + \int_0^t d\tau \, \partial_0^2 \, \varphi_\tau(t, \mathbf{x}) = g(t, \mathbf{x}) + \int_0^t d\tau \, \partial_0^2 \varphi_\tau(t, \mathbf{x}) \,. \tag{5.100}$$

If we add $-\Delta u = \int_0^t d\tau \, (-\Delta \varphi_\tau(t, \mathbf{x}))$ and take into account that φ_τ solves the wave equation we are left with

$$\Box u(t, \mathbf{x}) = g(t, \mathbf{x}) \,. \tag{5.101}$$

So as a function of t and $\mathbf{x}$ the integral

$$u(t, \mathbf{x}) = \int_0^t d\tau \, (t - \tau) \, M_{(t-\tau), \mathbf{x}}[g_\tau] \tag{5.102}$$

solves $\Box u = g$ with vanishing initial values $u(0, \mathbf{x}) = 0$ and $\partial_t u(0, \mathbf{x}) = 0$.

To evaluate the integral for $t > 0$ we substitute $\tau(r) = t - r$ and integrate over r,

$$u(t, \mathbf{x}) = \int_0^t dr \, r \, M_{r, \mathbf{x}}[g_{t-r}] = \frac{1}{4\pi} \int_0^t dr \, r^2 \int_{-1}^1 d\cos\theta \int_0^{2\pi} d\varphi \, \frac{1}{r} \, g(t - r, \mathbf{x} + r \, \mathbf{n}) \,. \tag{5.103}$$

The three integrals extend over the points $\mathbf{y} = \mathbf{x} + r \, \mathbf{n}$ of the ball $B_{t, \mathbf{x}}$ around $\mathbf{x}$ with radius t and $r = |\mathbf{x} - \mathbf{y}|$ is the distance of $\mathbf{y}$ to the center. So the solution of the inhomogeneous wave equation $\Box u = g$ with vanishing initial values is

$$t \geq 0 : \quad u(t, \mathbf{x}) = \frac{1}{4\pi} \int_{B_{t, \mathbf{x}}} d^3 y \, \frac{g(t - |\mathbf{x} - \mathbf{y}|, \mathbf{y})}{|\mathbf{x} - \mathbf{y}|} \,. \tag{5.104}$$

For $t < 0$ we substitute $\tau(r) = t + r$ in (5.102), then after the interchange of the lower and upper boundary the r-integral extends from 0 to $|t| = -t$ and the three integrals are again a volume integral over the ball $B_{t,\mathbf{x}}$ around $\mathbf{x}$ with radius $|t|$,

$$t \leq 0: \qquad u(t, \mathbf{x}) = \frac{1}{4\pi} \int\limits_{B_{t,\mathbf{x}}} d^3 y \frac{g(t + |\mathbf{x} - \mathbf{y}|, \mathbf{y})}{|\mathbf{x} - \mathbf{y}|}. \tag{5.105}$$

If more generally the initial values at $t = 0$ do not vanish, then the solution of the initial value problem of the inhomogeneous wave equation consists of this particular solution and a wave packet (5.92) with the initial values

$$u(t, \mathbf{x}) = t\, M_{t,\mathbf{x}}[\phi] + \partial_0 \left(t\, M_{t,\mathbf{x}}[\psi] \right) + \frac{1}{4\pi} \int\limits_{B_{t,\mathbf{x}}} d^3 y \frac{g(t - \operatorname{sign}(t)\, |\mathbf{x} - \mathbf{y}|, \mathbf{y})}{|\mathbf{x} - \mathbf{y}|}.$$

$$\tag{5.106}$$

The solution exists, is unique and depends continuously on the initial data, so the initial value problem of the inhomogeneous wave equation is well-posed.

The retarded potential

$$u_{\mathrm{ret}}(t, \mathbf{x}) = \frac{1}{4\pi} \int d^3 y \frac{g(t - |\mathbf{x} - \mathbf{y}|, \mathbf{y})}{|\mathbf{x} - \mathbf{y}|} \tag{5.107}$$

is a particular solution of $\Box u = g$ which corresponds to initial time $t = -\infty$. It vanishes at early times if g does. It may suffer infrared divergencies (i.e. the limit of increasing integration volume may not exist) if the sources g do not vanish sufficiently rapidly with increasing distance.

Poincaré Covariance of the Fields

The linear map of the source g to its retarded potential u_{ret} (5.107) is Poincaré invariant: if $x \mapsto x' = \Lambda x + a$ is a time orientation preserving Poincaré transformation (3.10), then the transformed source $\hat{g} : x \mapsto g(\Lambda^{-1}(x - a))$ (4.47) generates the transformed retarded potential $\hat{u}_{\mathrm{ret}} : x \mapsto u_{\mathrm{ret}}(\Lambda^{-1}(x - a))$.

To see this we shift in (5.107) the integration variable and integrate over $\mathbf{z} = \mathbf{x} - \mathbf{y}$

$$4\pi\, u_{\mathrm{ret}}(t, \mathbf{x}) = \int d^3 z \frac{g(t - |\mathbf{z}|, \mathbf{x} - \mathbf{z})}{|\mathbf{z}|}. \tag{5.108}$$

If we denote $(t, \mathbf{x})$ by x and $(|\mathbf{z}|, \mathbf{z})$ by z, then $f = 4\pi\, u_{\mathrm{ret}}$ is written as the special case $m^2 = 0$ of an integral

$$f(x) = \int \frac{d^3 z}{z^0} g(x - z), \quad z^0 = \sqrt{m^2 + \mathbf{z}^2}. \tag{5.109}$$

The integral extends over all points $x - z$ of the backward light cone of x, as seen from $z^2 = (z^0)^2 - \mathbf{z}^2 = 0$ and $z^0 > 0$. For $m^2 > 0$ the domain of integration, the points z with $z^2 = m^2 \geq 0, z^0 > 0$ constitute a mass shell.

The linear map of the source g to the potential f is Poincaré invariant. We will check that $\hat{f}(x) = f(\Lambda^{-1}(x - a))$ corresponds to $\hat{g}(x) = g(\Lambda^{-1}(x \quad a))$.

For translations, $x' = x + a$, $\Lambda = 1$, the translated potential

$$\hat{f}(x) = f(x - a) = \int \frac{d^3 z}{z^0} g(x - a - z) = \int \frac{d^3 z}{z^0} \hat{g}(x - z) \qquad (5.110)$$

is the potential of the translated source.

Because Lorentz transformations are linear, $\Lambda^{-1}x - z = \Lambda^{-1}(x - z')$ with $z' = \Lambda z$, one has

$$\hat{f}(x) = f(\Lambda^{-1}x) = \int \frac{d^3 z}{z^0} g(\Lambda^{-1}x - z)$$

$$= \int \frac{d^3 z}{z^0} g(\Lambda^{-1}(x - z')) = \int \frac{d^3 z}{z^0} \hat{g}(x - z') . \qquad (5.111)$$

If in particular Λ is a rotation or rotary reflection, $z'^0 = z^0, \mathbf{z}' = D\mathbf{z}$, then we can easily use rotated integration variables (z'^1, z'^2, z'^3). The modulus of the determinant of the Jacobi matrix $J = \partial z'/\partial z = D$ is $|\det D| = 1$ (6.6), and $z'^0 = \sqrt{m^2 + \mathbf{z}'^2}$ is unchanged by a rotation and coincides with $z^0 = \sqrt{m^2 + \mathbf{z}^2}$. So for rotations one concludes

$$\hat{f}(x) = \int \frac{d^3 z}{z^0} \hat{g}(x - z') = \int \frac{d^3 z'}{z'^0} \hat{g}(x - z') = \int \frac{d^3 z}{z^0} \hat{g}(x - z) . \qquad (5.112)$$

In the last step we have named the integration variable $\mathbf{z}$ again. So the rotated potential $\hat{u}_{\text{ret}}$ is the potential of the rotated source $\hat{g}$.

The measure $d^3 z/z^0$ is invariant under arbitrary time orientation preserving Lorentz transformations. We have to show this only for boosts in x-direction because each Lorentz transformation can be written as such a boost which is preceded and followed by a rotation (6.35).

The boost in x-direction (3.8), applied to the four-vector $(\sqrt{m^2 + \mathbf{z}^2}, \mathbf{z})$, gives

$$\sqrt{m^2 + \mathbf{z}'^2} = \frac{\sqrt{m^2 + \mathbf{z}^2} - v z_x}{\sqrt{1 - v^2}} = \frac{1}{\sqrt{1 - v^2}} \left(1 - \frac{v z_x}{\sqrt{m^2 + \mathbf{z}^2}}\right) \sqrt{m^2 + \mathbf{z}^2},$$

$$z'_x = \frac{z_x - v\sqrt{m^2 + \mathbf{z}^2}}{\sqrt{1 - v^2}}, \quad z'_y = z_y, \quad z'_z = z_z , \qquad (5.113)$$

the determinant of the 3×3 Jacobi matrix J, $J^i{}_j = \partial z'^i/\partial z^j$, is

$$\det J = \frac{\partial z'_x}{\partial z_x} = \frac{1}{\sqrt{1-v^2}}\left(1 - \frac{v\,z_x}{\sqrt{m^2+\mathbf{z}^2}}\right). \tag{5.114}$$

Altogether one has for each $m^2 \geq 0$

$$\frac{\mathrm{d}^3 z'}{\sqrt{m^2+\mathbf{z}'^2}} = \frac{\mathrm{d}^3 z}{\sqrt{m^2+\mathbf{z}'^2}}\,|\det J| = \frac{\mathrm{d}^3 z}{\sqrt{m^2+\mathbf{z}^2}}. \tag{5.115}$$

After changing and renaming the integration variable in (5.111) we can conclude

$$\hat{f}(x) = \int \frac{\mathrm{d}^3 z}{z^0}\,\hat{g}(x-z') = \int \frac{\mathrm{d}^3 z'}{z'^0}\,\hat{g}(x-z') = \int \frac{\mathrm{d}^3 z}{z^0}\,\hat{g}(x-z). \tag{5.116}$$

So the Lorentz transformed potential corresponds to the Lorentz transformed source. By the same reason the map of the current to the retarded four-potential

$$A^m_{\mathrm{ret}}(x) = \int \frac{\mathrm{d}^3 z}{|z^0|}\,j^m(x-z)\Big|_{z^0=\sqrt{\mathbf{z}^2}}, \tag{5.117}$$

is invariant under Poincaré transformations

$$\hat{A}^m(x) = \Lambda^m{}_n\,A^n(\Lambda^{-1}x - a),\quad \hat{j}^n(x) = \Lambda^n{}_m\,j^m(\Lambda^{-1}(x-a)). \tag{5.118}$$

The transformed current $\hat{j}$ is conserved conserved if j is,

$$\partial_{x^n}\hat{j}^n(x) = \Lambda^n{}_m\,\partial_{x^n}(\Lambda^{-1}(x-a))^r\,\partial_{z^r}j^m(z)\big|_{z=\Lambda^{-1}(x-a)}$$
$$= \Lambda^n{}_m\,\Lambda^{-1}{}^r{}_n\,\partial_{z^r}j^m(z) = \delta^r{}_m\,\partial_{z^r}j^m(z) = \partial_{z^m}j^m(z) = 0. \tag{5.119}$$

Also the map of the amplitudes to the wave packet (5.83) is Poincaré invariant under suitable transformations of the amplitudes,

$$A^n_{\mathrm{hom}}(x) = \int \tilde{\mathrm{d}}k\,\left(a^{n*}(k)\,\mathrm{e}^{\mathrm{i}k\cdot x} + a^n(k)\,\mathrm{e}^{-\mathrm{i}k\cdot x}\right)\Big|_{k^0=\sqrt{\mathbf{k}^2}}, \tag{5.120}$$

because the integration measure $\tilde{\mathrm{d}}k$ (5.84) and the domain of integration are both Poincaré invariant (5.115).

Under a translation $x \mapsto x + b$ the amplitudes $a = (a^0, a^1, a^2, a^3)$ change by multiplication with the function $f_b : k \mapsto \mathrm{e}^{\mathrm{i}k\cdot b}$, $\hat{a} = f_b\,a$ (a transformation which is not adjoint to the transformations N_g and M_g of the target and base manifold (4.47))

$$\hat{A}^n(x) = A^n(x-b) = \int \tilde{\mathrm{d}}k\,\left((a^{n*}(k)\,\mathrm{e}^{-\mathrm{i}k\cdot b})\,\mathrm{e}^{\mathrm{i}k\cdot x} + (a^n(k)\,\mathrm{e}^{\mathrm{i}k\cdot b})\,\mathrm{e}^{-\mathrm{i}k\cdot x}\right).$$
$$\tag{5.121}$$

Under Lorentz transformations, the amplitudes transform as a vector field, which is defined on the mass shell of momentum space.

To be specific, if we evaluate the transformed field $\hat{A}^n(x) = \Lambda^n{}_m A^m(\Lambda^{-1}x)$, then the scalar product $k \cdot (\Lambda^{-1}x)$ in (5.120) equals $k' \cdot x$ with $k' = \Lambda k$, because scalar products are Lorentz invariant. If we integrate over the three spatial components of $k' = \Lambda k$ rather than of k, use $a^m(k) = a^m(\Lambda^{-1}k')$ and the fact that $\tilde{d}k = \tilde{d}k'$ is a Lorentz invariant measure (5.115), then the transformed wave packet satisfies

$$\hat{A}^n(x) = \Lambda^n{}_m A^m(\Lambda^{-1}x) = \Lambda^n{}_m \int \tilde{d}k' \left(a^{m*}(\Lambda^{-1}k')\, e^{i k' \cdot x} + a^m(\Lambda^{-1}k')\, e^{-i k' \cdot x} \right).$$
(5.122)

The denomination of the integration variables, $\mathbf{k}$ or $\mathbf{k}'$, is irrelevant. So the Lorentz transformed wave packet corresponds to the Lorentz transformed amplitudes

$$\hat{a}^n(k) = \Lambda^n{}_m a^m(\Lambda^{-1}k).$$
(5.123)

The components of the wave packet are related by the Lorenz condition $\partial_m A^m = 0$ (5.72), which is already satisfied by the retarded potential,

$$\partial_m \int \frac{d^3 z}{|\mathbf{z}|}\, j^m(x - z) = \int \frac{d^3 z}{|\mathbf{z}|}\, \partial_m j^m(x - z) = 0.$$
(5.124)

Therefore the amplitudes $a^{n*}(k)$ of the wave packet are restricted

$$k_n a^{n*}(k) = 0.$$
(5.125)

Moreover, one can change them by a gauge transformation by the gradient of a function χ (5.68), which satisfies the homogeneous wave equation. Thereby the amplitudes change by

$$a'^{n*}(k) = a^{n*}(k) + i k^n c^*(k), \quad k^2 = 0,$$
(5.126)

where $c^*(k)$ is the amplitude of the gauge function χ.

By (5.125) only three of the four complex amplitudes of the four-potential are independent. The amplitude in the direction of k can be gauged away. So the wave packet contains for given wave vector $\mathbf{k}$ two degrees of freedom. They correspond to two independent transverse directions of polarization.

5.5 Action Principle and Noether's Theorems

Each field f, e.g. the field strength $\mathbf{E}$ or $\mathbf{B}$, is a differentiable map

$$f : \begin{cases} \mathcal{M} \to \mathcal{N} \\ x \mapsto f(x) \end{cases} \tag{5.127}$$

of a n-dimensional base space $\mathcal{M}$ to a d-dimensional target space $\mathcal{N}$. It defines a corresponding map, its lift $\hat{f}$ to the jet space $\mathcal{J}_k$,

$$\hat{f} : \begin{cases} \mathcal{M} \to \mathcal{J}_k \\ x \mapsto \hat{f}(x) = (x, f(x), \partial f(x), \ldots, \partial^k f(x)) \end{cases} \tag{5.128}$$

which maps each $x \in \mathcal{M}$ to x and the values of f and all its partial derivatives up to order k at this point.

We denote a typical point of $\mathcal{J}_k$ by its coordinates $(x, y, y_{(1)}, y_{(2)}, \ldots y_{(k)})$. Here x denotes a point in the base space, y is a d-tupel with components y^r, $y_{(1)}$ a point in the tangent space denoted by components y^r_m, where m ranges over n values. $y_{(l)}$ has components $y^r_{m_1 m_2 \ldots m_l}$ where each m_i ranges over n values. Because the order of partial derivatives can be exchanged, the components of the variables $y_{(l)}$ are completely symmetric under each permutation π of the derivative indices,

$$y^r_{m_1 m_2 \ldots m_l} = y^r_{m_{\pi(1)} m_{\pi(2)} \ldots m_{\pi(l)}}. \tag{5.129}$$

So there are $d \, (n+l)!/(n! \, l!)$ independent components of $y_{(l)}$. To be definite, one can order them lexicographically and restrict the independent ones by $m_1 \leq m_2 \ldots \leq m_l$.

We define the derivatives d_m of jet functions in analogy with (4.14)

$$d_m \phi = \left(\partial_{x^m} + \sum_{l \geq 0} \sum_r \sum_{m_1 \leq m_2 \ldots \leq m_l} y^r_{m \, m_1 m_2 \ldots m_l} \, \partial_{y^r_{m_1 m_2 \ldots m_l}} \right) \phi \tag{5.130}$$

such that their composition with the lift gives the derivatives of the composed function (4.15)

$$(d_m \phi) \circ \hat{f} = \partial_{x^m} (\phi \circ \hat{f}) . \tag{5.131}$$

A curve in the space of fields $\mathcal{F}$ is a one parameter family of fields $\lambda \mapsto f_\lambda$, where λ ranges in some interval. Jet functions ϕ, evaluated on the lift of f_λ, define by their change as a function of λ and by the chain rule the differential operator δ,

$$\frac{\partial}{\partial \lambda} (\phi \circ \hat{f}_\lambda) = (\delta \phi) \circ f_\lambda , \tag{5.132}$$

which acts on jet functions by

$$\delta = \delta y^r \partial_{y^r} + (d_m \delta y^r) \partial_{y^r_m} + \cdots \sum_r \sum_{m_1 \leq m_2 \ldots \leq m_k} (d_{m_1} \ldots d_{m_l} \delta y^r) \partial_{y^r_{m_1 m_2 \ldots m_l}} + \cdots , \tag{5.133}$$

where $\delta y \circ \hat{f}_\lambda = \frac{\partial}{\partial \lambda} f_\lambda$ is the change of f_λ and the derivatives of f_λ change by the derivatives of the change, $\delta y^r_{m_1 m_2 \ldots m_k} \circ \hat{f}_\lambda = (d_{m_1} d_{m_2} \ldots d_{m_k} \delta y^r) \circ \hat{f}_\lambda$, because derivatives with respect to coordinates and the parameter may be interchanged (4.18), i.e. d_m commutes with δ.

If there are curves through each point of $\mathscr{F}$ such that for all $f \in \mathscr{F}$ the change $\delta f = \delta y \circ \hat{f}$ is given by an d-tupel of jet functions δy, evaluated on the lift $\hat{f}$, we call δy and δ local (not to be confused with gauged).

The Lagrangian $\mathscr{L}$ is a function of some jet space, e.g. $\mathscr{J}_1$,

$$\mathscr{L} : \begin{cases} D \subset \mathbb{R}^{n+d+dn} \to \mathbb{R} \\ (x, y, y_{(1)}) \quad \mapsto \mathscr{L}(x, y, y_{(1)}) \end{cases} . \tag{5.134}$$

It defines a local functional on the space $\mathscr{F}$ of fields f, the action S,

$$S : \begin{cases} \mathscr{F} \to \mathbb{R} \\ f \mapsto S[f] = \int d^n x \, (\mathscr{L} \circ \hat{f})(x) = \int d^n x \, \mathscr{L}(x, f(x), \partial f(x)) \end{cases} . \tag{5.135}$$

The equations of motion state for the physical fields, that the action is stationary up to boundary terms, i.e. for all paths f_λ through $f_0 = f_{\text{phys}}$ the derivative of $S[f_\lambda]$ vanishes for $\lambda = 0$ up to boundary terms

$$0 = \frac{\partial}{\partial \lambda} S[f_\lambda]|_{\lambda=0} = \int d^n x \, (\delta \mathscr{L} \circ \hat{f}_{\text{phys}})(x) . \tag{5.136}$$

Because in $\delta \mathscr{L}$ the derivatives of δy can be shifted away

$$\delta \mathscr{L} = \delta y^r \partial_{y^r} \mathscr{L} + d_m (\delta y^r) \, \partial_{y^r_m} \mathscr{L} = \delta y^r \left(\partial_{y^r} \mathscr{L} - d_m \, \partial_{y^r_m} \mathscr{L} \right) + d_m (\delta y^r \, \partial_{y^r_m} \mathscr{L}) \tag{5.137}$$

at the cost of complete derivatives, which however only contribute boundary terms, and because the change of the action has to vanish for all $\delta f = \delta y \circ \hat{f}$ the Euler derivative of the Lagrangian

$$\frac{\hat{\partial} \mathscr{L}}{\hat{\partial} y^r} = \partial_{y^r} \mathscr{L} - d_m \partial_{y^r_m} \mathscr{L} \tag{5.138}$$

vanishes for physical fields, i.e. they satisfy the Euler-Lagrange equations

$$\frac{\hat{\partial} \mathscr{L}}{\hat{\partial} y} \circ \hat{f}_{\text{phys}} = 0 . \tag{5.139}$$

In particular the Maxwell equations (5.71) for the potential $A = (A_0, A_1, A_2, A_3)$ are the Euler-Lagrange equations of the action $S = S_{\text{Maxwell}} + S_{\text{matter}}$ with

$$S_{\text{Maxwell}}[A] = -\frac{1}{4} \int d^4 x \, F_{mn}(x) F^{mn}(x) , \quad F_{mn} = \partial_m A_n - \partial_n A_m . \tag{5.140}$$

The action of the charged matter provides by definition the current

$$\frac{\delta S_{\text{matter}}}{\delta A_m(x)} = -j^m(x) ,\tag{5.141}$$

which in the Maxwell equations is considered as given source, only restricted by the continuity equation $\partial_m j^m = 0$ (5.20).

Infinitesimal local transformations δy are maps of some jet space, e.g. $\mathscr{J}_1$, to the target space $\mathscr{N}$. They are called infinitesimal symmetries of the action S if the Lagrangian is invariant in first order up to derivative terms, i.e. if there exist jet functions K^m, such that

$$\delta y^r \, \partial_{y^r} \mathscr{L} + (d_m \delta y^r) \, \partial_{y^r_m} \mathscr{L} + d_m K^m = 0 \tag{5.142}$$

which is equivalent to

$$\delta y^r \frac{\hat{\partial} \mathscr{L}}{\hat{\partial} y^r} + d_m j^m = 0 \tag{5.143}$$

where the current j^m (and K^m to start with), given by

$$j^m = \delta y^r \, \partial_{y^r_m} \mathscr{L} + K^m + d_n B^{mn} , \tag{5.144}$$

is unique only up to so called improvement terms, i.e. jet functions $d_n B^{mn}$ with $B^{mn} = -B^{nm}$ [11, 26]. (5.143) is the

Noether Theorem of Field Theory: *To each infinitesimal symmetry δy of the action there corresponds a conserved current j. Vice versa to each conserved current there corresponds an infinitesimal symmetry of the action.*

By (5.143) the divergence of $j^m \circ \hat{f}_{\text{phys}}$ vanishes on account of the equations of motion (5.139). Conversely to each conserved current $\bar{j}$ there corresponds an infinitesimal symmetry δy of the action, because by definition jet functions $\bar{j}^m$ are the components of a conserved current if its divergence is proportional to (derivatives of) the equations of motion, i.e. if there are jet functions s^r and $s^{r\,m}$ such that

$$d_m \bar{j}^m + s^r \frac{\hat{\partial} \mathscr{L}}{\hat{\partial} y^r} + s^{r\,m} d_m \frac{\hat{\partial} \mathscr{L}}{\hat{\partial} y^r} = 0 , \tag{5.145}$$

$$d_m \left(\bar{j}^m + s^{r\,m} \frac{\hat{\partial} \mathscr{L}}{\hat{\partial} y^r} \right) + (s^r - d_m s^{r\,m}) \frac{\hat{\partial} \mathscr{L}}{\hat{\partial} y^r} = 0 . \tag{5.146}$$

This shows (5.143) with $j^m = \bar{j}^m + s^{r\,m} \frac{\hat{\partial} \mathscr{L}}{\hat{\partial} y^r}$ and $\delta y^r = (s^r - d_m s^{r\,m})$.

In particular, the infinitesimal translation of the potential A by a constant four-vector $-c$ accompanied by an infinitesimal gauge transformation (5.68) by the gradient of $-c^l A_l$

$$\delta A_n = c^l (d_l A_n - d_n A_l) = c^l F_{ln} \tag{5.147}$$

is a symmetry of the action S_{Maxwell} with the Lagrangian $\mathscr{L} = -1/4\, F_{mn}\, F^{mn}$. The Lagrangian changes by $\delta\mathscr{L} = -d_m(c^l F_{ln})F^{mn} = -1/2((d_m c^l F_{ln} - d_n c^l F_{lm})F^{mn}$, because the double sum with F^{mn} antisymmetrizes in m and n (5.27). Moreover, by construction of F as antisymmetrized derivatives of the potential A, the sum over the cyclic permutations of $d_m F_{ln}$ vanishes, therefore $\delta\mathscr{L} = -1/2(d_l c^l F_{mn})F^{mn} = d_l c^l \mathscr{L}$, i.e. δ is a symmetry (5.142) with $K^m = -c^m \mathscr{L}$.

The conserved current (5.144) turns out to be the energy-momentum tensor (5.23) in direction of c, $j^m = c_l T^{lm}$. This justifies to call $p^k = \int \mathrm{d}^3 x T^{k0}(x)$ (5.28) the energy and the momentum. They are the conserved quantities which correspond to the invariance of the action under translations in spacetime.

Because the energy-momentum tensor of the electromagnetic field is symmetric and traceless (5.24), the current $j^m = c_l T^{lm}$ is conserved not only for constant c but more generally if c satisfies the conformal Killing equation [25]

$$\partial_m c_n + \partial_n c_m - \frac{1}{2}\eta_{mn}\partial_k c^k = 0, \tag{5.148}$$

which has the general 15-parameter solution

$$c_m(x) = a_m + \omega_{mn} x^n + d\, x_m + b_m x \cdot x - 2(b \cdot x)x_m, \tag{5.149}$$

where a_m parameterizes an infinitesimal translation, $\omega_{mn} = -\omega_{nm}$ a Lorentz transformation, d a scale transformation and b_m a proper conformal transformation.

The action S_{Maxwell} of the potential A is invariant under infinitesimal conformal transformations $\delta A_n = c^m F_{mn}$ but the physical properties of matter are not invariant under dilations and proper conformal transformations: atoms exist only in their invariable size, not in each scaled version. That matter breaks the symmetry group of the electrodynamic interactions to the Poincaré group and not to a larger or smaller subgroup, e.g. that it does not define a rest frame, cannot be deduced logically from the Maxwell equations, but is the result of observations and experiments.

Exponentiated an infinitesimal proper conformal transformation yields the map

$$T_b : x \mapsto x' = \frac{x + b\, x^2}{1 + 2b \cdot x + b^2 x^2}. \tag{5.150}$$

This is not an invertible transformation of $\mathbb{R}^4$, no matter how small $b \neq 0$ is, because the denominator vanishes on the lightcone $(x + b/b^2)^2 = 0$ or, if $b^2 = 0$, on the plane $1 + 2b \cdot x = 0$. An infinitesimal symmetry, which corresponds to a conserved current, needs not be the derivative of a one parameter family of transformations.

If an infinitesimal symmetry δy contains an arbitrary function ξ of $\mathbb{R}^n$ and it's derivatives, i.e. if δy is a jet function also of a field coordinate ξ and it derivatives $\xi_{(1)}, \xi_{(2)} \ldots$, on which the Lagrangian does not depend, and is of the form

$$\delta y^r = \xi R^r + (d_m \xi)R^{r\,m}, \tag{5.151}$$

where R^r and $R^{r\,m}$ are some ξ-independent jet functions, then the transformation is called an infinitesimal gauge symmetry. For example, the change of the potential by the gradient of an arbitrary function ξ, $\delta A_m = d_m \xi$ (5.68), is a gauge symmetry of the action S_{Maxwell}, because already F_{mn} is invariant.

To each infinitesimal gauge symmetry there corresponds the Noether identity

$$\frac{\hat{\partial}}{\hat{\partial}\xi}\left(\delta y^r \frac{\hat{\partial}\mathscr{L}}{\hat{\partial}y^r}\right) = R^r \frac{\hat{\partial}\mathscr{L}}{\hat{\partial}y^r} - d_m\left(R^{r\,m}\frac{\hat{\partial}\mathscr{L}}{\hat{\partial}y^r}\right) = 0 \,, \qquad (5.152)$$

because in the Euler derivative of (5.143) with respect to ξ the Euler derivative of the derivatives $d_m j^m$ vanishes. Vice versa, to each identity

$$r^r \frac{\hat{\partial}\mathscr{L}}{\hat{\partial}y^r} + r^{r\,m} d_m \frac{\hat{\partial}\mathscr{L}}{\hat{\partial}y^r} = 0 \,, \qquad (5.153)$$

there corresponds the infinitesimal gauge symmetry $\delta y^r = \xi(r^r - d_m r^{r\,m}) - r^{r\,m} d_m \xi$. Just multiply the identity with ξ and rearrange the terms to obtain (5.143).

Second Noether Theorem: *To each infinitesimal gauge symmetry of the action corresponds an identity among the Euler derivatives of the Lagrangian. Vice versa, to each identity among the Euler derivatives of the Lagrangian there corresponds an infinitesimal gauge symmetry of the action.*

In electrodynamics the gauge transformation $\delta A_r = d_m \xi \delta^m{}_r$ is of the form (5.151) with $R_r = 0$ and $R_r{}^m = \delta^m{}_r$ and the Noether identity (5.152) states, that the divergence of the Euler derivative with respect to the potential vanishes, $d_m \frac{\hat{\partial}\mathscr{L}}{\hat{\partial}A_m} = 0$, in particular for $\mathscr{L}_{\text{Maxwell}}$ the identity is $d_m d_n F^{nm} = 0$.

If one inserts (5.151) into (5.143) and uses (5.152) one obtains for infinitesimal gauge symmetries of the Lagrangian $\mathscr{L}$

$$d_m(j^m + \xi R^{r\,m}\frac{\hat{\partial}\mathscr{L}}{\hat{\partial}y^r}) = 0 \qquad (5.154)$$

as identity in jet space. The divergence vanishes if and only if the terms in the braces are the divergence of antisymmetric jet functions [11, 26]

$$j^m = -\xi R^{r\,m}\frac{\hat{\partial}\mathscr{L}}{\hat{\partial}y^r} + d_n B^{nm}, \; B^{nm} = -B^{mn} \,. \qquad (5.155)$$

Therefore, in gauge theories by the equations of motion the charge in each volume is determined by a surface integral over its boundary, as $d_m B^{0m}$ does not contain a time derivative, but is the divergence of some three-vector field $\mathbf{E}$

$$Q_V = \int_V d^3x \, j^0 = \int_V d^3x \, \text{div}\mathbf{E} = \int_{\partial V} d^2\mathbf{f} \cdot \mathbf{E} \,. \tag{5.156}$$

Consider a gauge invariant matter Lagrangian $\mathscr{L}_{\text{matter}}$ with a fixed gauge field. Such a field is called a background field. It defines a subset of infinitesimal gauge transformations, called rigid transformations, which leave the background field invariant, $\delta A = 0$. For example, a vanishing potential $A = 0$ in electrodynamics is invariant under constant gauge transformations ξ, $\delta A = \partial \xi = 0$. As another example, the metric of Minkowski space is invariant under infinitesimal Poincaré transformations, which satisfy Killing's equation $\partial_m \xi_n + \partial_n \xi_m = 0$.

To these symmetries of the background there correspond conserved Noether currents which by (5.155) and by the equations of motion of the matter fields are given, up to improvement terms, by the Euler derivatives with respect to the gauge field, evaluated at the background field

$$j^m{}_{\text{rigid}} = \delta\phi_i \frac{\partial \mathscr{L}_{\text{matter}}}{\partial \partial_m \phi_i} + K^m + d_n B^{mn} = -\xi R^m_{A_k} \frac{\hat{\partial} \mathscr{L}_{\text{matter}}}{\hat{\partial} A_k} \,. \tag{5.157}$$

In general relativity, this explains why the Euler derivative of the matter Lagrangian with respect to the metric is the energy-momentum tensor $\mathscr{T}^{mn}$, the collection of Noether currents, which are conserved because Minkowski space is invariant under translations of space and time,

$$j^m \circ \hat{\phi}_{\text{phys}} \Big|_{g_{mn}=\eta_{mn}} \quad \mathscr{T}^{mn}\xi_n \,, \quad \mathscr{T}^{mn} = -\frac{1}{2} \frac{\hat{\partial} \mathscr{L}_{\text{matter}}}{\hat{\partial} g_{mn}} \,. \tag{5.158}$$

5.6 Charged Point Particle

A particle with charge q, which is forced to traverse the worldline $t \mapsto (t, \mathbf{z}(t))$, generates by its charge and current densities

$$\rho(t, \mathbf{y}) = q \, \delta^3(\mathbf{y} - \mathbf{z}(t)) \,, \quad \mathbf{j}(t, \mathbf{y}) = q \frac{d\mathbf{z}}{dt} \delta^3(\mathbf{y} - \mathbf{z}(t)) \,. \tag{5.159}$$

the scalar potential (5.117)

$$\phi(t, \mathbf{x}) = \frac{1}{4\pi} \int d^3y \frac{\rho(t_{\text{ret}}, \mathbf{y})}{|\mathbf{x} - \mathbf{y}|} = \frac{q}{4\pi} \int d^3y \frac{\delta^3(\mathbf{y} - \mathbf{z}(t_{\text{ret}}))}{|\mathbf{x} - \mathbf{y}|} \,, \quad t_{\text{ret}} = t - |\mathbf{x} - \mathbf{y}| \,. \tag{5.160}$$

The argument of the δ-function $\mathbf{y}' = \mathbf{y} - \mathbf{z}(t_{\text{ret}})$ is a composite function of the integration variable $\mathbf{y}$, as the retarded time $t_{\text{ret}} = t - |\mathbf{x} - \mathbf{y}|$ depends on $\mathbf{y}$.

By the substitution theorem for integrals such a composite δ-function, applied to a test function f, yields

$$\int d^3y\, \delta^3(y'(y))f(y) = \int d^3y' \left|\det \frac{\partial y}{\partial y'}\right| \delta^3(y')f(y(y'))$$

$$= \frac{1}{\left|\det \frac{\partial y'}{\partial y}\right|_{|\hat{y}}} f(\hat{y}), \quad y'(\hat{y}) = 0. \qquad (5.161)$$

In the case at hand, y' vanishes at $y = z(t_{\text{ret}})$. The test function f is the Coulomb potential $1/|x - z(t_{\text{ret}})|$. The Jacobi matrix of the substitution and its determinant is

$$\frac{\partial y'^i}{\partial y^j} = \delta^i{}_j + N^i{}_j\,, \ N^i{}_j = -\frac{dz^i}{dt}\frac{x^j - z^j}{|x - z|}\,, \det \frac{\partial y'}{\partial y} = 1 - \frac{dz}{dt}\frac{x - z}{|x - z|}\,, \quad (5.162)$$

where we exploited that N has rank 1 which implies $\det(1 + N) = 1 + \operatorname{tr} N$.

We obtain the scalar potential and analogously the vector potential

$$4\pi\phi(t, x) = \frac{q}{|x - z(t_{\text{ret}})| - \frac{dz}{dt}\frac{x-z}{|x-z|}}\,, \quad 4\pi A(t, x) = \frac{q}{|x - z(t_{\text{ret}})| - \frac{dz}{dt}\frac{x-z}{|x-z|}}\frac{dz}{dt}\,.$$
$$(5.163)$$

This four potential is named after Alfred-Marie Liénard and Emil Wiechert who derived it first. In the form

$$A^m(x) = \frac{q}{4\pi}\frac{u^m}{y \cdot u}\,, \qquad (5.164)$$

one can calculate the corresponding field strength with tolerable algebraic effort. $u = \frac{dz}{ds}$ is the normalized tangent vector to the worldline of the particle at its intersection $z(s(x))$ with the backward light cone of x. We parameterize the worldline with its proper time s, which is shown by a watch which is carried along.

$$u = \frac{dz}{ds} = \frac{1}{\sqrt{1 - v^2}}\binom{1}{v}\,, \quad v = \frac{dz}{dt}\,, \quad u^2 = 1\,. \qquad (5.165)$$

The lightlike vector $y = x - z(s(x))$ points from the cause to the effect: from the event $z(s(x))$, in which the particle crosses the backward light cone of x to the observer at x in direction n and in distance r

$$y = \binom{|x - z(s(x))|}{x - z(s(x))} = r\binom{1}{n}\,, \quad y^2 = 0\,, \quad y \cdot u = \frac{r}{\sqrt{1 - v^2}}(1 - v\,n)\,. \quad (5.166)$$

The proper time s on the worldline defines the time $s(x)$, which an observer at x reads from the clock. It has the constant value s on the forward light cone of $z(s)$.

The gradient k_m of $s(x)$ is calculated be differentiation of $y^2 = 0$,

$$0 = (\delta_m{}^n - u^n \partial_m s)\, y_n, \quad k_m := \partial_m s = \frac{y_m}{y \cdot u}. \tag{5.167}$$

It is lightlike $k^2 = 0$ and given by

$$k = \frac{\sqrt{1-v^2}}{1-vn}\begin{pmatrix} 1 \\ n \end{pmatrix} \tag{5.168}$$

With its help we can express the derivatives of y and $y \cdot u$ by the four-acceleration

$$\dot{u} = \frac{du}{ds} = \frac{dt}{ds}\frac{d}{dt}\left(\frac{1}{\sqrt{1-v^2}}\begin{pmatrix} 1 \\ v \end{pmatrix}\right)$$

$$= \frac{1}{(1-v^2)^2}\begin{pmatrix} v\,a \\ a\,(1-v^2) + v\,(v\,a) \end{pmatrix}, \quad a = \frac{d^2 z}{dt^2} \tag{5.169}$$

and the quantities which we have introduced,

$$\partial_m y^n = \delta_m{}^n - u^n k_m\,,$$

$$\partial_m (y \cdot u) = (\partial_m y^n)\, u_n + y \cdot \dot{u}\, k_m = u_m + (y \cdot \dot{u} - 1)\, k_m\,. \tag{5.170}$$

We obtain the field strengths

$$F_{mn} = \partial_m A_n - \partial_n A_m = -\frac{q}{4\pi\,(y \cdot u)^2}\partial_m(y \cdot u)\, u_n + \partial_m u_n \frac{q}{4\pi\,(y \cdot u)} - m \leftrightarrow n$$

$$= k_m w_n - k_n w_m\,, \quad w_m = \frac{q}{4\pi\,(y \cdot u)^2}u_m + \frac{q}{4\pi\,(y \cdot u)}(\dot{u}_m - u_m\, k \cdot \dot{u})\,. \tag{5.171}$$

and in particular the electric field, $E^i = F_{0i} = k_0 w_i - k_i w_0$,

$$\mathbf{E}(t, \mathbf{x}) = \frac{q\,(1-v^2)}{4\pi\,r^2\,(1-\mathbf{v}\,\mathbf{n})^3}(\mathbf{n} - \mathbf{v}) + \frac{q}{4\pi\,r\,(1-\mathbf{v}\,\mathbf{n})^3}\mathbf{n} \times ((\mathbf{n} - \mathbf{v}) \times \mathbf{a})\,. \tag{5.172}$$

The part which does not depend on the acceleration, decreases with distance as $1/r^2$ and does not show in the direction $\mathbf{n}$ from the cause, the event z, to the effect at x, but into the direction $\mathbf{x} - (\mathbf{z} + r\mathbf{v})$, away from the destination $\mathbf{z} + r\mathbf{v}$, which the particle would reach with constant velocity $\mathbf{v}$ in the moment when it effects the field at $\mathbf{x}$.

The acceleration dependent part, the radiation field, decreases with $1/r$ and is orthogonal to the direction $\mathbf{n}$ from the cause to its effect at $\mathbf{x}$.

The magnetic field of the particle is orthogonal to $\mathbf{n}$ and $\mathbf{E}$

$$B^k = -\varepsilon_{ijk} k_i w_j = \varepsilon_{ijk} k^i w_j = \varepsilon_{ijk} k^i/k^0\, E^j, \quad \mathbf{B} = \mathbf{n} \times \mathbf{E}\,. \tag{5.173}$$

The energy current density $\mathbf{S} = \mathbf{E} \times \mathbf{B}$ of the radiation points into the direction $\mathbf{n}$ away from the cause: an accelerated charge looses energy by radiation.

Mathematically and physically the description of matter by point particles is untenable. If no interaction contributes with a negative energy density, then the charge of the electron cannot be concentrated in a sphere with a smaller radius than half the electron radius

$$r_{electron} = \frac{e^2}{4\pi \, \varepsilon_0 \, m_{electron} c^2} = 2,818 \cdot 10^{-15} \, \text{m} \,. \qquad (5.174)$$

Otherwise the electric field outside of an electron at rest would contain already more energy than the rest energy of the electron. Extended charge distributions require interactions which prevent the charge distributions of the particles to fly apart. The charge distribution cannot be rigid, because in relativistic physics there cannot exist truly rigid bodies. Therefore, matter has to be described by a field theory. According to our present understanding, only quantum field theory yields an acceptable theory for matter.

Chapter 6
The Lorentz Group

Abstract Rotations in higher dimensional spaces define one- and two dimensional subspaces in which they act just as in the Euclidean plane. Similarly Lorentz transformations in higher dimensional spaces act, up to rotations, in two dimensional subspaces as boosts. Each Lorentz transformation Λ of the four-dimensional spacetime corresponds uniquely to a pair $\pm M$ of linear transformations of a complex two-dimensional space, the space of spinors. Their inspection reveals that aberration, the Lorentz transformation of the directions of light rays, acts as a Möbius transformation of the Riemann sphere.

6.1 Rotations

Rotations and rotary reflections of a d-dimensional Euclidean space $\mathscr{V}$ [1]

$$D : \begin{cases} \mathscr{V} & \to \mathscr{V} \\ (v^1, v^2, \ldots v^d) & \mapsto (D^1{}_j v^j, D^2{}_j v^j, \ldots D^d{}_j v^j) \end{cases} \tag{6.1}$$

constitute the subgroup $O(d)$ of linear transformations, which leave the scalar product (here given in an orthonormal basis)

$$u \cdot v = u^j \delta_{jk} v^k = u^1 v^1 + u^2 v^2 + \cdots + u^d v^d \tag{6.2}$$

$$\delta_{jk} : \begin{cases} 1 \text{ if } j = k \\ 0 \text{ if } j \neq k \end{cases} \tag{6.3}$$

of all vectors $v = (v^1, v^2, \ldots v^d)$ and $u = (u^1, u^2, \ldots u^d)$ and thereby all angles and lengths invariant,

---

[1] We use Einstein's summation convention. Unless stated otherwise, each pair of indices denotes the sum over the range of its values, $D^i{}_j v^j = D^i{}_1 v^1 + D^i{}_2 v^2 + \cdots + D^i{}_d v^d$.

N. Dragon, *The Geometry of Special Relativity—a Concise Course*,
SpringerBriefs in Physics, DOI: 10.1007/978-3-642-28329-1_6,
© The Author(s) 2012

$$D^i{}_j u^j D^i{}_k v^k = u^j \delta_{jk} v^k. \tag{6.4}$$

This holds if and only if the orthogonality relations hold,

$$D^i{}_j D^i{}_k = \delta_{jk}. \tag{6.5}$$

The columns of the matrix D contain the components of orthonormal vectors, or phrased as a matrix property, the transposed matrix D^T, $D^T{}_j{}^i = D^i{}_j$, equals the inverse matrix,

$$D^T D = 1. \tag{6.6}$$

In particular $(\det D)^2 = 1$, because of $1 = \det 1 = \det(D^T D) = \det D^T \det D = (\det D)^2$. The determinant of D therefore is 1 or -1.

Rotary reflections and rotations form the group $O(d)$ of the orthogonal transformations in d dimensions. We reserve the name rotations for the orientation preserving transformations with determinant $\det D = 1$. They constitute the group of special orthogonal transformations, $SO(d)$.

The determinant is a continuous function of the matrix elements. Therefore there does not exist a one parameter set of rotary reflections D_λ, which varies continuously with λ and connects a rotation with $\det D_{\lambda=0} = 1$ and a reflection with $\det D_{\lambda=1} = -1$: rotations are not connected to reflections.

Each rotation can be generated by repeated application of infinitesimal rotations, it is continuously connected to the identity 1. In other words, the group $SO(d)$ is connected. This is seen by inspecting the eigenvalue equation of rotary reflections,

$$D w = \lambda w \quad w \neq 0. \tag{6.7}$$

The eigenvalues λ satisfy the characteristic equation

$$\det(D - \lambda 1) = 0. \tag{6.8}$$

For real $d \times d$-matrices this is a polynomial equation of order d with real coefficients and has d not necessarily distinct complex solutions, where we count each real solution as complex, though special, solution.

To each complex eigenvalue $\lambda = \sigma + i\tau$ of the real rotary reflection $D = D^*$ there belongs an eigenvector with complex components, $w^i = u^i + i v^i$. The conjugate $\lambda^* = \sigma - i\tau$ is eigenvalue of the eigenvector $w^{*i} = u^i - i v^i$. For $D = D^*$ is real and conjugation yields $0 = ((D - \lambda 1)^i{}_j w^j)^* = (D - \lambda^* 1)^i{}_j w^{*j}$.

By the orthogonality relations (6.5) and the eigenvalue equation

$$(D w^*)^i (D w)^i - w^{*i} w^i = 0 = (\lambda^* \lambda - 1) w^{*i} w^i = (|\lambda|^2 - 1)(u^2 + v^2)$$
$$(D w)^i (D w)^i - w^i w^i = 0 = (\lambda^2 - 1) w^{*i} w^i = (\lambda^2 - 1)(u^2 - v^2 + 2 i u \cdot v), \tag{6.9}$$

and because $u^2 + v^2$ does not vanish, each eigenvalue of an orthogonal transformation has modulus $|\lambda|^2 = 1$. Each real eigenvalue λ is 1 or -1. The corresponding eigenvector is invariant $Dn = n$, or is reflected, $Da = -a$.

If the eigenvalue $\lambda = \cos\alpha + i\sin\alpha$ is not real, then $(\lambda^2 - 1)$ does not vanish and u and v are orthogonal and have equal length. We choose the normalized vectors $e_1 = v/|v|$ and $e_2 = u/|u|$ as basis and spell out the eigenvalue equation

$$D(e_2 + ie_1) = (\cos\alpha + i\sin\alpha)(e_2 + ie_1) \tag{6.10}$$

or, decomposed into real- and imaginary part,

$$\begin{aligned} De_1 &= (\cos\alpha)\,e_1 + (\sin\alpha)\,e_2 \,, \\ De_2 &= -(\sin\alpha)\,e_1 + (\cos\alpha)\,e_2 \,, \end{aligned} \tag{6.11}$$

i.e. on this orthonormal basis of the plane which is spanned by e_1 and e_2 the rotary reflection acts by the matrix (3.6)

$$D_\alpha = \begin{pmatrix} \cos\alpha & -\sin\alpha \\ \sin\alpha & \cos\alpha \end{pmatrix}. \tag{6.12}$$

We can restrict α to the range $0 < \alpha < \pi$, i.e. $\sin\alpha > 0$, because a negative sign of $\sin\alpha$ can be absorbed by choosing the basis $e_1' = e_1, e_2' = -e_2$.

The real subspace $\mathscr{U}_\perp$ of vectors y, which are orthogonal to a real or complex eigenvector w, $y \cdot w = 0$, is mapped to itself, $D(\mathscr{U}_\perp) \subset \mathscr{U}_\perp$,

$$y \cdot w = 0 = (Dy) \cdot (Dw) = \lambda(Dy) \cdot w \quad DU_\perp \subset U_\perp. \tag{6.13}$$

The space $\mathscr{U}_\perp$ together with e_1 and e_2 or n or a spans the Euclidean space $\mathscr{V}$ because, due to the positive definiteness of the scalar product, the vectors in $\mathscr{U}_\perp$ are linearly independent from e_1 and e_2 or n or a.

Restricted to $\mathscr{U}_\perp$ the rotation D has an eigenvalue which is either real, then D leaves the eigenvector invariant or reflects it, or the eigenvalue is not real, in which case D acts by a rotation D_β of the two dimensional eigenplane. Altogether, there is a orthonormal basis of $\mathscr{V}$ in which the matrix D takes the form

$$D = \begin{pmatrix} D_\alpha & & & & \\ & \ddots & & & \\ & & D_\beta & & \\ & & & 1 & \\ & & & & -1 \end{pmatrix}, \tag{6.14}$$

where $\mathbf{1}$ denotes a block of eigenvalues 1 and $-\mathbf{1}$ another block of eigenvalues -1. If the dimension of $\mathscr{V}$ is odd, there has to exist a real eigenvalue 1 or -1, there is an axis of rotation or of a reflection.

In particular, in $d = 3$ dimensions we can use the axis $\mathbf{n}$ to decompose each vector into its parallel and orthogonal parts, $\mathbf{k} = \mathbf{k}_\parallel + \mathbf{k}_\perp$, $\mathbf{k}_\parallel = \mathbf{n}\,(\mathbf{n}\cdot\mathbf{k})$. They are completed by $\mathbf{n}\times\mathbf{k}_\perp$ to a right handed basis (if $\mathbf{k}_\perp$ does not vanish). In these terms the rotation by an angle α is given by

$$D_{\alpha\,\mathbf{n}}(\mathbf{k}_\parallel + \mathbf{k}_\perp) = \mathbf{k}_\parallel + (\cos\alpha)\,\mathbf{k}_\perp + (\sin\alpha)\,\mathbf{n}\times\mathbf{k}_\perp. \qquad (6.15)$$

In case of a proper reflection the number of eigenvalues -1 is odd, in case of a rotation even. Each pair of eigenvalues -1 can be considered to belong to a rotation D_π by $\alpha = \pi$.

Each rotation D, $\det D = 1$, is an iterated infinitesimal rotation $D = \mathrm{e}^\delta$. For eigenvectors n to the eigenvalue 1 we define $\delta n = 0$ and get $\mathrm{e}^\delta n = \delta^0 n = n$. For the orthonormal basis e_1 and e_2 which corresponds to D_α we define

$$\delta e_1 = \alpha\, e_2 \quad \delta e_2 = -\alpha\, e_1 \qquad (6.16)$$

consequently $\delta^2 e_1 = -\alpha^2 e_1$ and $\delta^2 e_2 = -\alpha^2 e_2$. Inserted into the exponential series e^δ and separated into even and odd powers

$$\mathrm{e}^\delta = \sum_k \frac{1}{(2k)!}\delta^{2k} + \sum_k \frac{1}{(2k+1)!}\delta^{2k+1} = \sum_k \frac{(-1)^k}{(2k)!}\alpha^{2k} + \sum_k \frac{(-1)^k}{(2k+1)!}\alpha^{2k}\delta$$

$$(6.17)$$

and applied to e_1 and e_2 one obtains (6.11)

$$\mathrm{e}^\delta e_1 = (\cos\alpha)\,e_1 + (\sin\alpha)\,e_2 \quad \mathrm{e}^\delta e_2 = -(\sin\alpha)\,e_1 + (\cos\alpha)\,e_2. \qquad (6.18)$$

Therefore $D = \mathrm{e}^\delta$ for the vectors $e_{1,i}$ and $e_{2,i}$, which are rotated by an angle α_i. Vectors a_k which are reflected, $Da_k = -a_k$ occur in pairs, because $\det D = 1$. Their transformation is a rotation by $\alpha = \pi$ and is also of the form $D = \mathrm{e}^\delta$. So this equation holds on a basis and therefore in all the Euclidean space.

Because each angle of rotation can be continuously increased from zero to α, all rotations are continuously connected to $\mathbf{1}$ and to one another. Consequently the orthogonal group $\mathrm{O}(d)$ consists of two disconnected components, the group of rotations $\mathrm{SO}(d)$ and the set $\mathscr{P}\,\mathrm{SO}(d)$, where the parity transformation $\mathscr{P}$ reflects an odd dimensional subspace, $\mathscr{P}a = -a$, and leaves the orthogonal subspace pointwise invariant, $\mathscr{P}n = n$.

6.2 Lorentz Transformations

The Lorentz group $\mathrm{O}(p,q)$, $p > 0$, $q > 0$, consists of the real, linear transformations Λ of the points x of the $p + q$-dimensional Minkowski space $\mathbb{R}^{p,q}$,

$$x' = \Lambda x \qquad (6.19)$$

which leave invariant the scalar product of $\mathbb{R}^{p,q}$ (here given in an orthonormal basis)

$$x \cdot y = \sum_{i=1}^{p} x^i y^i - \sum_{i=p+1}^{p+q} x^i y^i. \tag{6.20}$$

In matrix notation it is given by $x \cdot y = x^T \eta \, y$ with a matrix

$$\eta = \begin{pmatrix} 1 & \\ & -1 \end{pmatrix} \tag{6.21}$$

which has a $p \times p$ unit block and a $q \times q$ block -1. So Lorentz matrices have to satisfy $(\Lambda x)^T \eta \, \Lambda y = x^T \eta \, y$ for all x and y and therefore

$$\Lambda^T \eta \Lambda = \eta. \tag{6.22}$$

The determinant of this matrix equation leads to

$$(\det \Lambda)^2 = 1 \tag{6.23}$$

because $\det(\Lambda^T \eta \Lambda) = (\det \Lambda^T)(\det \eta)(\det \Lambda)$ and because $\det \Lambda^T = \det \Lambda$. So the determinant of a Lorentz transformation can only have the values $+1$ or -1. The special transformations Λ with $\det \Lambda = 1$ constitute the special orthogonal group $SO(p, q)$.

The group $O(p, q)$ is the manifold $O(p) \times O(q) \times \mathbb{R}^{pq}$ and consists of four disconnected components.

To show this, we decompose the $(p+q) \times (p+q)$-matrix Λ into a $p \times p$-matrix A, a $q \times q$-matrix D, a $q \times p$-matrix C and a $p \times q$-matrix B

$$\Lambda = \begin{pmatrix} A & B \\ C & D \end{pmatrix} \tag{6.24}$$

and write (6.22) in terms of these matrices,

$$A^T A = 1 + C^T C, \quad D^T D = 1 + B^T B, \quad A^T B = C^T D. \tag{6.25}$$

The symmetric matrix $C^T C$ is diagonalizable by a rotation and has nonnegative diagonal elements $\lambda_j = \sum_i C_{ij} C_{ij}$. Therefore the eigenvalues of $1 + C^T C$ are not smaller than 1 and A is invertible, $(\det A)^2 = \det(1 + C^T C) \geq 1$. The same applies to D, $(\det D)^2 \geq 1$.

An invertible real matrix A can be uniquely decomposed into the product of an orthogonal transformation O, $O^T = O^{-1}$, and a symmetric matrix S, $S^T = S$, with positive eigenvalues,

$$A = OS. \tag{6.26}$$

For $A^T A$ defines a symmetric matrix S^2 with positive eigenvalues $\lambda_i > 0$, $i = 1, \ldots p$, and also the positive symmetric matrix

$$S = \sqrt{A^T A}, \quad S = S^T, \tag{6.27}$$

with the same eigenvectors as S^2 and the positive eigenvalues $\sqrt{\lambda_i}$. The matrix

$$O = AS^{-1}, \quad O^T = O^{-1}, \tag{6.28}$$

is orthogonal as $S^{T-1}A^T A S^{-1} = S^{-1}S^2 S^{-1} = 1$ shows.

As A and D are invertible, Λ (6.24) can be decomposed uniquely (6.26)

$$\Lambda = \begin{pmatrix} O & \\ & \hat{O} \end{pmatrix} \begin{pmatrix} S & Q \\ P & \hat{S} \end{pmatrix}, \tag{6.29}$$

with $P = \hat{O}^{-1}C$ and $Q = O^{-1}B$. S and $\hat{S}$ are invertible and symmetric, O and $\hat{O}$ are orthogonal matrices. Equation (6.22) reads

$$S^2 = 1 + P^T P \quad SQ = P^T \hat{S} \quad \hat{S}^2 = 1 + Q^T Q. \tag{6.30}$$

Using $Q = S^{-1}P^T \hat{S}$ and $S^{-2} = (1 + P^T P)^{-1}$ in the last equation, one obtains

$$\hat{S}^2 = 1 + \hat{S}PS^{-1}S^{-1}P^T \hat{S} \quad \text{or} \quad 1 = \hat{S}^{-2} + P(1 + P^T P)^{-1}P^T. \tag{6.31}$$

Its consequences for $\hat{S}^{-2}$ are exhibited by applying the matrices to eigenvectors w of PP^T, $PP^T w = \lambda w$. If $P^T w$ is not zero, then it is eigenvector of $P^T P$, $(P^T P)P^T w = \lambda P^T w$, with the same eigenvalue. This is why

$$P(1 + P^T P)^{-1}P^T w = P\frac{1}{1+\lambda}P^T w = \frac{\lambda}{1+\lambda}w. \tag{6.32}$$

The equation applies also if $P^T w$ vanishes as then $PP^T w = 0$ which is $\lambda = 0$. Inserted into (6.31) one obtains $\frac{1}{1+\lambda}w = \hat{S}^{-2}w$ or $\hat{S}^2 w = (1+\lambda)w$. The eigenvectors w constitute a basis of $\mathbb{R}^q$. Therefore the equation holds for each vector, i.e. as matrix equation,

$$\hat{S}^2 = 1 + PP^T. \tag{6.33}$$

By the same reasons one concludes

$$Q = S^{-1}P^T \hat{S} = \sqrt{1 + P^T P}^{-1}P^T \sqrt{1 + PP^T} = P^T. \tag{6.34}$$

for eigenvectors of PP^T and therefore as matrix equation.

So each Lorentz matrix is uniquely given by a pair of orthogonal transformations $O \in O(p)$, $\hat{O} \in O(q)$ and a rotation-free Lorentz transformation L_P, $L_P = (L_P)^T$, which is determined by a matrix P with q rows and p columns

$$\Lambda = \begin{pmatrix} O & \\ & \hat{O} \end{pmatrix} L_P, \quad L_P = \begin{pmatrix} \sqrt{1 + P^T P} & P^T \\ P & \sqrt{1 + P P^T} \end{pmatrix}, \quad (L_P)^{-1} = L_{-P}.$$

(6.35)

Such $q \times p$ matrices P constitute the vector space $\mathbb{R}^{qp}$. Each Lorentz transformation corresponds one to one to a point in the manifold $O(p) \times O(q) \times \mathbb{R}^{qp}$.

The vector space $\mathbb{R}^{qp}$ is connected, the orthogonal groups consist of two disconnected components, therefore $O(p, q)$ has four disconnected components.

Lorentz transformations Λ with $\det O = \det \hat{O} = 1$ preserve the orientation of the timelike and spacelike directions and constitute the proper Lorentz group $SO(p, q)^\uparrow$. It is connected. The other components of $O(p, q)$ are obtained by multiplication with the inversion of time $\mathcal{T}$ and by the parity transformation $\mathcal{P}$, which reflect an odd dimensional timelike or spacelike subspace, and by $\mathcal{T}\mathcal{P}$,

$$\mathcal{T} = \begin{pmatrix} -1 & & & \\ & 1 & & \\ & & \ddots & \\ & & & 1 \end{pmatrix}, \quad \mathcal{P} = \begin{pmatrix} 1 & & & \\ & \ddots & & \\ & & 1 & \\ & & & -1 \end{pmatrix}.$$

(6.36)

The Lorentz transformation L_P acts in $1 + 1$-dimensional subspaces U_i, which are mutually orthogonal, of the Minkowski space $\mathbb{R}^{p,q}$ as the transformation (3.4). The subspace which is orthogonal to all of the U_i is pointwise invariant.

This follows from the inspection of eigenvectors w_i of PP^T, which acts on the subspace $\mathbb{R}^q$ with a definite scalar product. They are orthogonal to each other if they correspond to different eigenvalues λ_i and can be chosen to be mutually orthogonal, if the eigenvalue is degenerate. We choose them to be normalized, $w_j \cdot w_i = -w_j^T w_i = -\delta_{ij}$.

The eigenvectors u of PP^T with vanishing eigenvalue, $PP^T u = 0$, are annihilated already by P^T, $P^T u = 0$, as $PP^T u = 0$ implies $u^T PP^T u = 0$. This is a sum of squares of the components of $P^T u$ and vanishes only if $P^T u = 0$. Therefore L_P leaves each u invariant, $L_P u = u$.

Similarly one has $Pv = 0$ and $L_P v = v$ for each eigenvector $v \in \mathbb{R}^p$ of $P^T P$ with vanishing eigenvalue.

Each normalized eigenvector e_1 of PP^T with nonvanishing eigenvalue λ defines an orthogonal, normalized eigenvector $e_0 = P^T e_1/\sqrt{\lambda}$ of $P^T P$ with the same eigenvalue. On these vectors L_P acts by

$$L_P(e_0) = \sqrt{1 + \lambda}\, e_0 + \sqrt{\lambda}\, e_1, \quad L_P(e_1) = \sqrt{\lambda}\, e_0 + \sqrt{1 + \lambda}\, e_1$$

(6.37)

and is given in this basis by the matrix

$$\begin{pmatrix} \sqrt{1 + \lambda} & \sqrt{\lambda} \\ \sqrt{\lambda} & \sqrt{1 + \lambda} \end{pmatrix}.$$

(6.38)

This is the two-dimensional Lorentz boost (3.4) with velocity $v = -\sqrt{\lambda/(1+\lambda)}$.

Though each L_p decomposes in an appropriate basis into a sum of two-dimensional boosts, it is not true that each Lorentz transformation Λ can be decomposed into the transformations of two-dimensional subspaces if $(p-1)(q-1) \geq 1$. As a counter example consider the indecomposable Jordan blocks

$$J_\lambda = \begin{pmatrix} \lambda & 1 \\ & \lambda \end{pmatrix}. \tag{6.39}$$

The Lorentz transformation, which consists of such 2×2-blocks,

$$\Lambda = \begin{pmatrix} J_\lambda & \\ & J_{\lambda^{-1}} \end{pmatrix} \tag{6.40}$$

leaves invariant the scalar product

$$\eta = \begin{pmatrix} & A \\ A^T & \end{pmatrix}, \ A = \begin{pmatrix} & \lambda \\ -\lambda^{-1} & \end{pmatrix} \tag{6.41}$$

which in a different basis is diagonal with two timelike and two spacelike directions, $\eta \sim (1, 1, -1, -1)$.

In the four-dimensional spacetime $\mathbb{R}^{1,3}$ the rotation-free Lorentz transformation (6.35) has the form

$$L_p = \frac{1}{m} \begin{pmatrix} p^0 & p^j \\ p^i & m\delta^{ij} + \frac{p^i p^j}{p^0+m} \end{pmatrix}, \quad p^0 = \sqrt{m^2 + \mathbf{p}^2}. \tag{6.42}$$

Here we denote the components of the 3×1 column matrix P by p^i/m, $i = 1, 2, 3$, the spatial columns are enumerated by j, $j = 1, 2, 3$. The coefficients of δ^{ij} and $p^i p^j$ are determined by the requirement that the 3×3-matrix, in which they appear, has to act on eigenvectors of PP^T as the matrix $\sqrt{1 + PP^T}$. As P is just a column vector $\mathbf{p}$, all vectors orthogonal to it are eigenvectors of PP^T with eigenvalue 0 and P itself is eigenvector with eigenvalue $\lambda = \mathbf{p}^2/m^2$.

The rotation-free Lorentz transformation L_p transforms the four-momentum $\underline{p}$ of a particle with mass m at rest, $\underline{p} = (m, 0, 0, 0)$, into the four-momentum $p = (p^0, p^1, p^2, p^3)$ of a particle which moves with a velocity $v^i = p^i/p^0$ (3.47),

$$L_p \, \underline{p} = p. \tag{6.43}$$

6.3 The Rotation Group SU(2)/$\mathbb{Z}_2$

The unitary, unimodular transformations of a two-dimensional complex vector space constitute the group SU(2). As manifold it is the three sphere S^3.

The unitarity condition $U^\dagger U = \mathbf{1}$ states that the columns of each unitary 2×2-matrix U contain the components of an orthonormal basis. If U has the form

$$U = \begin{pmatrix} a & c \\ b & d \end{pmatrix}, \tag{6.44}$$

then $|a|^2 + |b|^2 = 1$. The second column is orthogonal to the first, $a^*c + b^*d = 0$, if and only if (c, d) is a multiple of $(-b^*, a^*)$ and the condition $\det U = 1$ determines this multiple

$$U = \begin{pmatrix} a & -b^* \\ b & a^* \end{pmatrix}, \quad (\Re a)^2 + (\Im a)^2 + (\Re b)^2 + (\Im b)^2 = 1. \tag{6.45}$$

To each U corresponds a point on $S^3 = \{(v, x, y, z) \in \mathbb{R}^4 : v^2 + x^2 + y^2 + z^2 = 1\}$ and to each point on S^3 there corresponds a $U \in \mathrm{SU}(2)$ with $a = v + \mathrm{i}z$ and $b = x + \mathrm{i}y$.

Points on S^3 can be designated by an angle $0 \le \alpha \le 2\pi$ and a three-dimensional unit vector (n_x, n_y, n_z) as $(v, x, y, z) = \cos \alpha/2\,(1, 0, 0, 0) - \sin \alpha/2\,(0, n_x, n_y, n_z)$. Correspondingly one can write each SU(2)-matrix as the following linear combination of the **1**-matrix σ^0 and the three Pauli matrices σ^i, $i = 1, 2, 3$,

$$\sigma^0 = \begin{pmatrix} 1 & \\ & 1 \end{pmatrix} \quad \sigma^1 = \begin{pmatrix} & 1 \\ 1 & \end{pmatrix} \quad \sigma^2 = \begin{pmatrix} & -\mathrm{i} \\ \mathrm{i} & \end{pmatrix} \quad \sigma^3 = \begin{pmatrix} 1 & \\ & -1 \end{pmatrix}, \tag{6.46}$$

$$U = \left(\cos \frac{\alpha}{2}\right) \mathbf{1} - \mathrm{i} \left(\sin \frac{\alpha}{2}\right) \mathbf{n}\cdot\sigma = \begin{pmatrix} \cos \frac{\alpha}{2} - \mathrm{i}\,(\sin \frac{\alpha}{2})\, n_z & -\mathrm{i}\,(\sin \frac{\alpha}{2})\,(n_x - \mathrm{i}n_y) \\ -\mathrm{i}\,(\sin \frac{\alpha}{2})\,(n_x + \mathrm{i}n_y) & \cos \frac{\alpha}{2} + \mathrm{i}\,(\sin \frac{\alpha}{2})\, n_z \end{pmatrix}. \tag{6.47}$$

Elementary calculation confirms the nine products of the Pauli matrices,

$$\sigma^i \sigma^j = \delta^{ij} \mathbf{1} + \mathrm{i}\varepsilon_{ijk}\sigma^k \quad i, j, k \in \{1, 2, 3\}. \tag{6.48}$$

If multiplied and summed with m^i and n^j this can also be written as

$$(\mathbf{m} \cdot \sigma)(\mathbf{n} \cdot \sigma) = (\mathbf{m} \cdot \mathbf{n})\mathbf{1} + \mathrm{i}(\mathbf{m} \times \mathbf{n}) \cdot \sigma. \tag{6.49}$$

In particular for each unit vector $\mathbf{n}$ the square simplifies $(\mathbf{n} \cdot \sigma)^2 = \mathbf{1}$ and with $(\mathbf{n} \cdot \sigma)^{2k} = \mathbf{1}$ and $(\mathbf{n} \cdot \sigma)^{2k+1} = \mathbf{n} \cdot \sigma$ power series of $\mathbf{n} \cdot \sigma$ simplify

$$\exp\left(-i\frac{\alpha}{2}\,\mathbf{n}\cdot\sigma\right) = \sum_k \frac{(-i\alpha/2)^{2k}}{(2k)!}\,\mathbf{1} + \sum_k \frac{(-i\alpha/2)^{2k+1}}{(2k+1)!}\,\mathbf{n}\cdot\sigma$$

$$= \sum_k \frac{(-1)^k(\frac{\alpha}{2})^{2k}}{(2k)!}\,\mathbf{1} - i\sum_k \frac{(-1)^k(\frac{\alpha}{2})^{2k+1}}{(2k+1)!}\,\mathbf{n}\cdot\sigma$$

$$= \left(\cos\frac{\alpha}{2}\right)\mathbf{1} - i\left(\sin\frac{\alpha}{2}\right)\mathbf{n}\cdot\sigma\,. \tag{6.50}$$

So each $U \in \mathrm{SU}(2)$ is generated by an infinitesimal transformation $-i\frac{\alpha}{2}\,\mathbf{n}\cdot\sigma$,

$$U = \exp\left(-i\frac{\alpha}{2}\,\mathbf{n}\cdot\sigma\right) = \cos\frac{\alpha}{2}\,\mathbf{1} - i\sin\frac{\alpha}{2}\,\mathbf{n}\cdot\sigma\,. \tag{6.51}$$

Each subgroup H of a group G defines an equivalence relation among its elements by

$$g \overset{H}{\sim} g' \Leftrightarrow g^{-1}g' \in H\,. \tag{6.52}$$

The set of equivalence classes is denoted by G/H. If moreover $ghg^{-1} \in H$ for all $g \in G$ and all $h \in H$, then the subgroup H is called normal and the equivalence classes form a group G/H, because the product g_2g_1 is equivalent to the product of equivalent elements $g_2h_2g_1h_1 = g_2g_1(g_1^{-1}h_2g_1)h_1 = g_2g_1h'$.

For instance, $\mathbb{Z}_2$, the cyclic group with two elements ± 1, is a normal subgroup of $\mathrm{SU}(2)$. Matrices $U \in \mathrm{SU}(2)$ are $\mathbb{Z}_2$-equivalent, if they differ at most by their sign. The group $\mathrm{SU}(2)/\mathbb{Z}_2$ consists out of $\mathrm{SU}(2)$-transformations up to the sign.

The group $\mathrm{SO}(3)$ of rotations in three dimensions is isomorphic to the group $\mathrm{SU}(2)/\mathbb{Z}_2$. This is to say, there exists a bijective, i.e. invertible and exhaustive, map D from $\mathrm{SU}(2)/\mathbb{Z}_2$ to $\mathrm{SO}(3)$, which maps each pair $\pm U$ of unitary 2×2-matrices with $\det U = 1$ to a 3×3-rotation $D_U = D_{-U}$ and which is compatible with the group multiplication, $D_{U_1 U_2} = D_{U_1} D_{U_2}$. The map is exhaustive, each rotation $R \in \mathrm{SO}(3)$ is a representation D_U of some unitary 2×2-matrix U. The inverse image of D_U in $\mathrm{SU}(2)/\mathbb{Z}_2$ is unique. If $D_U = D_V$ then $U = V$ or $U = -V$.

The rotation $D_U \in \mathrm{SO}(3)$ which corresponds to $U \in \mathrm{SU}(2)$ is the linear map

$$D_U : K \mapsto K' = UKU^\dagger \tag{6.53}$$

of hermitean, traceless 2×2-matrices K to hermitean traceless matrices K'. A 2×2-matric K is hermitean, $K = K^\dagger$,

$$\begin{pmatrix} k^{11} & k^{12} \\ k^{21} & k^{22} \end{pmatrix} = \begin{pmatrix} k^{11} & k^{12} \\ k^{21} & k^{22} \end{pmatrix}^\dagger = \begin{pmatrix} k^{11*} & k^{21*} \\ k^{12*} & k^{22*} \end{pmatrix}, \tag{6.54}$$

if the matrix elements k^{11} and k^{22} are real and if k^{12} is the complex conjugate of k^{21}. The matrix is traceless if $k^{11} = -k^{22}$. Such matrices span a three-dimensional real vector space and can be written as real linear combinations of the Pauli matrices (6.46)

$$K = \mathbf{k} \cdot \sigma = k^i \sigma^i = \begin{pmatrix} k^3 & k^1 - ik^2 \\ k^1 + ik^2 & -k^3 \end{pmatrix}. \tag{6.55}$$

The linear transformation $\mathbf{k} \cdot \sigma \mapsto U(\mathbf{k} \cdot \sigma)U^\dagger = \mathbf{k}' \cdot \sigma$ is a rotation $\mathbf{k} \mapsto \mathbf{k}' = D_U \mathbf{k}$. To check this by explicit calculation, we write (6.51) with $c = \cos \frac{\alpha}{2}$ and $s = \sin \frac{\alpha}{2}$ just as $U = c - is\mathbf{n} \cdot \sigma$ and decompose $\mathbf{k} = k_\parallel \mathbf{n} + \mathbf{k}_\perp$ into a part which is parallel to $\mathbf{n}$ and a transverse part. $K = K_\parallel + K_\perp$ with $K_\parallel = k_\parallel \mathbf{n} \cdot \sigma$ and $K_\perp = \mathbf{k}_\perp \cdot \sigma$.

The part $K_\parallel$ commutes with each power series in $\mathbf{n} \cdot \sigma$, in particular with U, and therefore is invariant

$$U K_\parallel U^\dagger = K_\parallel U U^\dagger = K_\parallel. \tag{6.56}$$

In the calculation of the transformation of $\mathbf{k}_\perp \cdot \sigma$ we observe that $\mathbf{k}_\perp \sigma$ anticommutes with $\mathbf{n} \cdot \sigma$ (6.49)

$$(\mathbf{k}_\perp \cdot \sigma)(\mathbf{n} \cdot \sigma) = -(\mathbf{n} \cdot \sigma)(\mathbf{k}_\perp \cdot \sigma) \tag{6.57}$$

because $\mathbf{k}_\perp$ and $\mathbf{n}$ are orthogonal to each other. Together with $(\mathbf{n} \cdot \sigma)^2 = 1$ and (6.49) we obtain

$$\begin{aligned} U(\mathbf{k}_\perp \cdot \sigma)U^\dagger &= (c - is\mathbf{n} \cdot \sigma)(\mathbf{k}_\perp \cdot \sigma)(c + is\mathbf{n} \cdot \sigma) \\ &= (c - is\mathbf{n} \cdot \sigma)(c - is\mathbf{n} \cdot \sigma)(\mathbf{k}_\perp \cdot \sigma) \\ &= U^2(\mathbf{k}_\perp \cdot \sigma) = (\cos\alpha - i\sin\alpha \, \mathbf{n} \cdot \sigma)(\mathbf{k}_\perp \cdot \sigma) \\ &= (\cos\alpha \, \mathbf{k}_\perp + \sin\alpha \, \mathbf{n} \times \mathbf{k}_\perp) \cdot \sigma. \end{aligned} \tag{6.58}$$

So $U = e^{-i\frac{\alpha}{2}\mathbf{n} \cdot \sigma}$ causes by $\mathbf{k} \cdot \sigma \mapsto U\mathbf{k} \cdot \sigma U^\dagger = (D_U \mathbf{k}) \cdot \sigma$ the rotation D_U of vectors $\mathbf{k}$ around the axis $\mathbf{n}$ by the angle α (6.15)

$$D_U : k_\parallel + \mathbf{k}_\perp \mapsto k_\parallel + \cos\alpha \, \mathbf{k}_\perp + \sin\alpha \, \mathbf{n} \times \mathbf{k}_\perp. \tag{6.59}$$

Vice versa, there corresponds to each rotation $D_{\alpha \mathbf{n}}$ around the axis $\mathbf{n}$ by the angle α the pair of unitary matrices $U = e^{-i\frac{\alpha}{2}\mathbf{n} \cdot \sigma}$ and $-U = e^{-i\frac{\alpha+2\pi}{2}\mathbf{n} \cdot \sigma}$.

6.4 The Group SL(2, $\mathbb{C}$)

Each invertible, complex matrix M can be uniquely decomposed into a product of a unitary matrix U, $U^\dagger = U^{-1}$, and the exponential e^H (which has positive eigenvalues) of a hermitean matrix, $H = H^\dagger$,

$$M = U e^H. \tag{6.60}$$

This equation generalizes the well known decomposition $z = e^{i\varphi} r$ of a nonvanishing complex number z into its phase and its modulus $r > 0$.

The polar decomposition exists because

$$e^{2H} = M^\dagger M \tag{6.61}$$

defines uniquely a hermitean matrix $H = H^\dagger$ because $M^\dagger M$ is hermitean, and therefore diagonalizable and has positive eigenvalues $\lambda = e^{2h}$. Therefore the hermitean matrix $H = \frac{1}{2} \ln M^\dagger M$ exists which has the same eigenvectors as $M^\dagger M$ and the real eigenvalues h. The matrix

$$U = Me^{-H} \tag{6.62}$$

is unitary, $(Me^{-H})^\dagger (Me^{-H}) = e^{-H} M^\dagger M e^{-H} = e^{-H} e^{2H} e^{-H} = 1$, and $M = Ue^H$.

As each invertible complex matrix M corresponds one to one to a pair (U, H) of a unitary and a hermitean matrix and because the hermitean $N \times N$-matrices constitute a real N^2-dimensional vector space, therefore the group $GL(N, \mathbb{C})$ of the general linear transformations in N complex dimensions is the manifold $U(N) \times \mathbb{R}^{N^2}$.

If the matrix M belongs to the subgroup $SL(N, \mathbb{C})$ of the special linear transformations with $\det M = 1$, then the trace of H, the sum of its diagonal elements, vanishes, $\operatorname{tr} H = 0$, because $\det(e^{2H}) = \det(M^\dagger M) = 1$ and $\det(e^{2H})$ is the product of its eigenvalues e^{2h}. Moreover U is unimodular, $\det U = \det(Me^{-H}) = 1$. The group $SL(N, \mathbb{C})$ therefore is the manifold $SU(N) \times \mathbb{R}^{(N^2-1)}$. In particular, $SU(2)$ is the manifold S^3 (6.45) and $SL(2, \mathbb{C})$ is the manifold $S^3 \times \mathbb{R}^3$.

The unique decomposition of each Lorentz transformation into an orthogonal transformation and a rotation-free Lorentz transformation (6.35) shows that the group $SO(1, 3)^\uparrow$ of the transformations which preserve the time orientation and the space orientation is the manifold $SO(3) \times \mathbb{R}^3$. The rotation group $SO(3)$ is $S^3/\mathbb{Z}_2$. Therefore $SL(2, \mathbb{C})$ is the covering manifold of $SO(1, 3)^\uparrow$.

Also as group $SL(2, \mathbb{C})$ is a covering of $SO(1, 3)^\uparrow$. This means: there is a four-dimensional, real representation of $SL(2, \mathbb{C})$, which maps each complex 2×2-matrix M with $\det M = 1$ to a Lorentz transformation Λ_M with $\Lambda^0{}_0 \geq 1$ and $\det \Lambda = 1$ and which is compatible with the group product, $\Lambda_{M_1 M_2} = \Lambda_{M_1} \Lambda_{M_2}$. The representation is exhaustive, each Lorentz transformation Λ with $\Lambda^0{}_0 \geq 1$ and $\det \Lambda = 1$ can be written as representation Λ_M of a complex matrix M with $\det M = 1$. The preimage of Λ is unique up to the sign. One has $\Lambda_M = \Lambda_N$ exactly if $M = N$ or $M = -N$.

The transformation Λ_M is the linear map

$$\Lambda_M : \hat{k} \mapsto \hat{k}' = M \hat{k} M^\dagger \tag{6.63}$$

of hermitean 2×2-matrices $\hat{k} = \hat{k}^\dagger$ to hermitean matrices $\hat{k}'$. They are real linear combinations

$$\hat{k} = k^0\sigma^0 - k^1\sigma^1 - k^2\sigma^2 - k^3\sigma^3 = \begin{pmatrix} k^0 & -k^3 & -k^1 + ik^2 \\ -k^1 - ik^2 & k^0 + k^3 \end{pmatrix} \qquad (6.64)$$

of the matrix $\sigma^0 = \mathbf{1}$ and the three Pauli matrices (6.46), so they span a four-dimensional real vector space. Also $M\,\hat{k}\,M^\dagger$ is hermitean and defines a real four-vector $k' = (k'^0, k'^1, k'^2, k'^3)$. Its components

$$k'^m = \Lambda^m{}_n k^n \qquad (6.65)$$

are linear in k with real matrix elements $\Lambda^m{}_n$.

The matrices Λ are a representation of the group SL(2, $\mathbb{C}$), as $\Lambda_{M_1 M_2} = \Lambda_{M_1} \Lambda_{M_2}$ holds for successive transformations

$$\hat{k}'' = M_1 M_2 \hat{k} M_2^\dagger M_1^\dagger$$
$$k''^m = (\Lambda_{M_1})^m{}_r k'^r = (\Lambda_{M_1})^m{}_r (\Lambda_{M_2})^r{}_n k^n = (\Lambda_{M_1 M_2})^m{}_n k^n. \qquad (6.66)$$

The transformation Λ_M is a Lorentz transformation, because the determinant of the 2×2-matrix $\hat{k}$ is a quadratic polynomial, namely the length squared (2.46) of the four-vector k,

$$\det \hat{k} = (k^0)^2 - (k^1)^2 - (k^2)^2 - (k^3)^2 \qquad (6.67)$$

and coincides because of $\det M = 1$ with the determinant of $\hat{k}' = M\,\hat{k}\,M^\dagger$. Therefore $k'^2 = k^2$ and $k' = \Lambda k$ is a Lorentz transformation.

We have already shown, that each $U \in$ SU(2) can be written as $e^{-i\frac{\alpha}{2}\mathbf{n}\cdot\sigma}$ (6.51) and causes, just as $-U$, a rotation of $\mathbf{k}$ around the axis $\mathbf{n}$ by the angle α (6.59).

We calculate the Lorentz transformation which corresponds to the hermitean factor e^H in $M = Ue^H$. The traceless, hermitean matrix H is a linear combination $H = -\frac{\beta}{2}\mathbf{n}\cdot\sigma$ of the three Pauli matrices (6.46) ($\mathbf{n}$ denotes a unit vector). The exponential series e^H simplifies due to $(\mathbf{n}\cdot\sigma)^2 = \mathbf{1}$ as in (6.50)

$$e^H = \exp\left(-\frac{\beta}{2}\mathbf{n}\cdot\sigma\right) = \left(\cosh\frac{\beta}{2}\right) - \left(\sinh\frac{\beta}{2}\right)\mathbf{n}\cdot\sigma \qquad (6.68)$$

(we do not write the $\mathbf{1}$-matrix explicitly).

The matrix $\hat{k}$, which is mapped to $\hat{k}' = e^H \hat{k} (e^H)^\dagger = e^H \hat{k} e^H$, is decomposed into $\hat{k} = \hat{k}_\| + \hat{k}_\perp$, where $\hat{k}_\| = k^0 - k_\|\mathbf{n}\cdot\sigma$ and $\hat{k}_\perp = -\mathbf{k}_\perp\cdot\sigma$ and where $\mathbf{k} = k_\|\mathbf{n} + \mathbf{k}_\perp$ decomposes the vector into its parts which are parallel and orthogonal to $\mathbf{n}$.

For the calculation of the transformation of $\hat{k}_\perp$ we only need the fact that $\hat{k}_\perp$ anticommutes with $H \propto \mathbf{n}\cdot\sigma$ (6.57), $\hat{k}_\perp H = -H\hat{k}_\perp$, because $\mathbf{k}_\perp$ and $\mathbf{n}$ are orthogonal to each other,

$$e^H \hat{k}_\perp e^H = e^H e^{-H} \hat{k}_\perp = \hat{k}_\perp. \qquad (6.69)$$

The perpendicular part $\mathbf{k}_\perp$ is unchanged.

The matrix $\hat{k}_\parallel$ commutes with e^H. Moreover $(\mathbf{n} \cdot \sigma)^2 = 1$ implies

$$
\begin{aligned}
e^H \hat{k}_\parallel e^H = \hat{k}_\parallel e^{2H} &= (k^0 - k_\parallel \mathbf{n} \cdot \sigma)(\cosh \beta - (\sinh \beta) \mathbf{n} \cdot \sigma) \\
&= \big((\cosh \beta) k^0 + (\sinh \beta) k_\parallel\big) - \big((\sinh \beta) k^0 + (\cosh \beta) k_\parallel\big) \mathbf{n} \cdot \sigma \quad (6.70) \\
&= k'^0 - k'_\parallel \mathbf{n} \cdot \sigma.
\end{aligned}
$$

We conclude

$$
\begin{pmatrix} k'^0 \\ k'_\parallel \end{pmatrix} = \begin{pmatrix} \cosh \beta & \sinh \beta \\ \sinh \beta & \cosh \beta \end{pmatrix} \begin{pmatrix} k^0 \\ k_\parallel \end{pmatrix}. \tag{6.71}
$$

This is the rotation-free Lorentz transformation, the boost, in direction of $\mathbf{n}$ with velocity $v = \tanh\beta$. Up to the sign of v the transformation coincides with (3.4).

In contrast to rotations and rotation-free Lorentz transformations, not all matrices $M \in \mathrm{SL}(2, \mathbb{C})$ can be written as exponential of an infinitesimal transformation

$$
N = \exp((\mathbf{k} + i\mathbf{l}) \cdot \sigma) = \cosh z + \frac{\sinh z}{z}(\mathbf{k} + i\mathbf{l}) \cdot \sigma \quad (\mathbf{k} + i\mathbf{l})^2 = z^2. \tag{6.72}
$$

There are exceptions of the form

$$
M = \begin{pmatrix} -1 & b \\ 0 & -1 \end{pmatrix}, \quad b \neq 0. \tag{6.73}
$$

$\frac{\sinh z}{z}$ must not vanish, otherwise N would be diagonal could not be M. For the main diagonal elements to coincide, one has to have $k^3 = l^3 = 0$. $N_{12} = 0$ states $k^1 + il^1 + i(k^2 + il^2) = 0$. Consequently $z = 0$ and $N_{11} = \cosh z = 1 \neq M_{11}$.

The matrix M is not an exponential but a product $M = U e^H$ (6.60) of exponentials. This shows that the Hausdorff-series for $C(A, B)$ in $e^A e^B = e^C$ has only a restricted domain of convergence.

6.5 Möbius Transformations of Light Rays

The determinant of $\hat{k}$ (6.64) vanishes if k is the the wave-vector of a light ray, because k is lightlike and $\det \hat{k} = k^2 = 0$ (6.67). As the matrix $\hat{k}$ only has rank 1, its elements are products of the components of a two-dimensional, complex vector χ, which is unique up to a phase $e^{i\gamma}$, with the components of the conjugate vector,

$$
\begin{pmatrix} k^0 - k^3 & -k^1 + ik^2 \\ -k^1 - ik^2 & k^0 + k^3 \end{pmatrix} = \begin{pmatrix} \chi_1 \\ \chi_2 \end{pmatrix} (\chi_1^*, \chi_2^*), \quad \begin{pmatrix} \chi_1 \\ \chi_2 \end{pmatrix} = e^{i\gamma} \begin{pmatrix} \sqrt{k^0 - k^3} \\ -\dfrac{k^1 + ik^2}{\sqrt{k^0 - k^3}} \end{pmatrix}.
\tag{6.74}
$$

Lorentz transformations change $\hat{k} = \chi \chi^\dagger$ into $k' = M \hat{k} M^\dagger = (M\chi)(M\chi)^\dagger$, i.e. χ transforms as a two-dimensional complex vector into $\chi' = M \chi$. A vector with this SL(2, $\mathbb{C}$) transformation is called spinor,

$$\begin{pmatrix} \chi'_1 \\ \chi'_2 \end{pmatrix} = \begin{pmatrix} a\chi_1 + b\chi_2 \\ c\chi_1 + d\chi_2 \end{pmatrix} \quad M = \begin{pmatrix} a & b \\ c & d \end{pmatrix}, \quad a, b, c, d \in \mathbb{C} \quad ad - bc = 1 . \tag{6.75}$$

The ratio $z = \chi_1/\chi_2$ corresponds one to one to the incident direction $\mathbf{e}$, from which the light ray is seen. The wave vector has the form $(|\mathbf{k}|, \mathbf{k}) = |\mathbf{k}|(1, -\mathbf{e})$. In spherical coordinates $\mathbf{e} = (\sin\theta \cos\varphi, \sin\theta \sin\varphi, \cos\theta)$ (2.34) and using the identity (3.18) one has

$$z = \frac{\chi_1}{\chi_2} = -\frac{k^0 - k^3}{k^1 + ik^2} = \frac{1 + \cos\theta}{\sin\theta \, e^{i\varphi}} = \cot\frac{\theta}{2} \, e^{-i\varphi} . \tag{6.76}$$

The set of directions is the Riemann sphere $\mathbb{C} \cup \{\infty\}$.

As the direction z of a light ray is the ratio of spinor components, it is changed by Lorentz transformations Λ, which correspond to the pair $\pm M(\Lambda) \in$ SL(2, $\mathbb{C}$)/$\mathbb{Z}_2$, by the corresponding Möbius transformation

$$T_M : z \mapsto \frac{az + b}{cz + d} . \tag{6.77}$$

Aberration and rotation are Möbius transformations of $z = \cot\frac{\theta}{2} e^{-i\varphi}$.

Two Möbius transformations T_M and T_N coincide if $M = N$ or $M = -N$, i.e. the Möbius group is isomorphic to SL(2, $\mathbb{C}$)/$\mathbb{Z}_2$ and therefore to the proper Lorentz group SO(1, 3)†.

If z_1, z_2, z_3 are three different points on the Riemann sphere $\mathbb{C} \cup \{\infty\}$ and if also w_1, w_2, w_3 are different, then there is exactly one Möbius transformation [22]

$$z \mapsto Tz : \quad \frac{(Tz - w_1)(w_2 - w_3)}{(Tz - w_2)(w_1 - w_3)} = \frac{(z - z_1)(z_2 - z_3)}{(z - z_2)(z_1 - z_3)} , \tag{6.78}$$

which maps z_1 to $w_1 = Tz_1$, z_2 to $w_2 = Tz_2$ and z_3 to $w_3 = Tz_3$.

Therefore, at a given place there is exactly one observer, who perceives three given stars in three given directions. The positions of the other stars are then fixed.

References

1. V.I. Arnold, *Mathematical Methods of Classical Mechanics* (Springer, Berlin, 1980)
2. A. Aspect, J. Dalibard, G. Roger, Experimental test of Bell's in-equalities using time-varying analyzers. Phys. Rev. Lett. **49**, 1804–1807 (1982)
3. J. Bailey et al., Il Nuovo Cimento **9A**, 369 (1972)
4. J.S. Bell, On the Einstein–Podolsky–Rosen paradox, Physics **1**, 195–200 (1964), reprinted in [5]
5. J.S. Bell, *Speakable and Unspeakable in Quantum Mechanics* (Cambridge University Press, Cambridge, 1987)
6. M.V. Berry, Regular and irregular motion, in *Topics in Nonlinear Dynamics*. Amer. Inst. Phys. Conf. Proceedings Nr, ed. by S. Jorna, vol. 46, (1978) p. 16
7. H. Bondi, *Relativity and Common Sense* (Heinemann, London, 1965)
8. I. Ciufolini, J.A. Wheeler, Gravitation and Inertia, Princeton Series in Physics (Princeton University Press, Princeton, 1995)
9. J.F. Clausner, M.A. Horne, A. Shimony, R.A. Holt, Proposed experiment to test local hidden-variable theories. Phys. Rev. Lett **23**, 880–884 (1969)
10. R. Courant, D. Hilbert, *Methods of Mathematical Physics II* (Wiley, Chichester, 1953)
11. N. Dragon, *BRST Cohomology*, http://www.itp.uni-hannover.de/dragon
12. N. Dragon, N. Mokros, *Relativistic Flight through Stonehenge* (1999), http://www.itp.uni-hannover.de/dragon
13. C.W.F. Everitt et al., Gravity probe B: final results of a space experiment to test general relativity. Phys. Rev. Lett. **106**, 221–101 (2011)
14. J.C. Hafele, R.E. Keating, Around-the-world atomic clocks: predicted relativistic time gains. Science **177**, 166–167 (1972): Around-the-world atomic clocks: observed relativistic time values, Science **177**, 168–170 (1972)
15. R.A. d'Inverno, *Introducing Einstein's Relativity* (Oxford University Press, Oxford, 1992)
16. J.D. Jackson, L.B. Okun, Historical roots of gauge invariance. Rev. Mod. Phys. **73**, 663–680 (2001)
17. R. Kippenhahn, *Light from the Depths of Time* (Springer, Berlin, 1987)
18. A. Lampa, Wie erscheint nach der Relativitätstheorie ein bewegter Stab einem ruhenden Beobachter? Zeitschrift for Physik **27**, 138–148 (1924)
19. D.-E. Liebscher, *Einstein's Relativity and the Geometries of the Plane* (Wiley-VCH Verlag GmbH, Germany, 1998)
20. L.V. Lorenz, On the identity of the vibrations of light with electrical currents. Phil. Mag. Ser. **4**(34), 287–301 (1867)

N. Dragon, *The Geometry of Special Relativity—a Concise Course*,
SpringerBriefs in Physics, DOI: 10.1007/978-3-642-28329-1,
© The Author(s) 2012

21. J. Moser, *Stable and Random Motion in Dynamical Systems* (Princeton University Press, Princeton, 1973)
22. T. Needham, *Visual Complex Analysis* (Clarendon Press, Oxford, 1997)
23. P. Nemec, http://www.ohg-sb.de/lehrer/nemec/relativ.htmwww.ohg-sb.de/lehrer/nemec/relativ.htm
24. E. Noether, Invariante Variationsprobleme, Nachrichten von der Königlichen Gesellschaft der Wissenschaften zu Göttingen, Mathematisch-physikalische Klasse, pp. 235 – 257 (1918)
25. J. O'Connor, E. Robertson, *The MacTutor History of Mathematics Archive*, http://www-groups.dcs.st-and.ac.uk/history/BiogIndex.html
26. P.J. Olver, *Applications of Lie Groups to Differential Equations* (Springer, Berlin, 1986)
27. B. Parkinson, J. Spilker (eds.), *Global Positioning System: Theory and Applications*, vol. I. (American Institute of Aeronautics and Astronautics, Washington, 1996)
28. Particle Data Group, K. Nakamura et al., J. Phys. G 37 075021 (2010), http://pdg.lbl.gov
29. R. Penrose, The apparent shape of a relativistically moving sphere. Proc. Camb. Phil. Soc. **55**, 137–139 (1959)
30. G. Seeber, *Satellite Geodesy: Foundations, Methods and Applications* (de Gruyther, New York, 1993)
31. L. Stodolsky, The speed of light and the speed of neutrinos. Phys. Lett. B **201**, 353 (1988)
32. J.L. Synge, *Relativity: The General Theory* (North-Holland, Amsterdam, 1964) (The name Synge is pronounced sing)
33. J. Terrell, Invisibility of Lorentz Contraction. Phys. Rev. **116**, 1041–1045 (1959)
34. C.M. Will, *Theory and Experiment in Gravitational Physics* (Cambridge University Press, Cambridge, 1993)
35. C.M. Will, *The Confrontation between General Relativity and Experiment*, http://arxiv.org/abs/gr-qc/0103036
36. http://mathworld.wolfram.com/Hyperboloid.html
37. M.J.W. Nicholas, *Special Relativity*, Lecture Notes in Physics m6 (Springer, Berlin, 1992)

Index

N. Dragon, *The Geometry of Special Relativity—a Concise Course*,
SpringerBriefs in Physics, DOI: 10.1007/978-3-642-28329-1,
© The Author(s) 2012